Series on Complexity Science – Vol. 2

Stochastic Dynamics of Complex Systems

From Glasses to Evolution

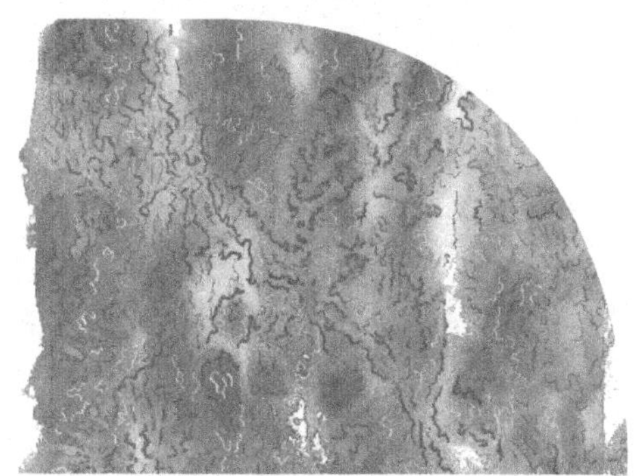

Series on Complexity Science

ISSN: 1755-7453

Series Editor: Henrik Jeldtoft Jensen *(Imperial College London, UK)*

Published

Vol. 1: A Complexity Approach to Sustainability: Theory and Application
by Angela Espinosa & Jon Walker

Vol. 2: Stochastic Dynamics of Complex Systems: From Glasses to Evolution
by Paolo Sibani & Henrik Jeldtoft Jensen

Series on Complexity Science – Vol. 2

Stochastic Dynamics of Complex Systems

From Glasses to Evolution

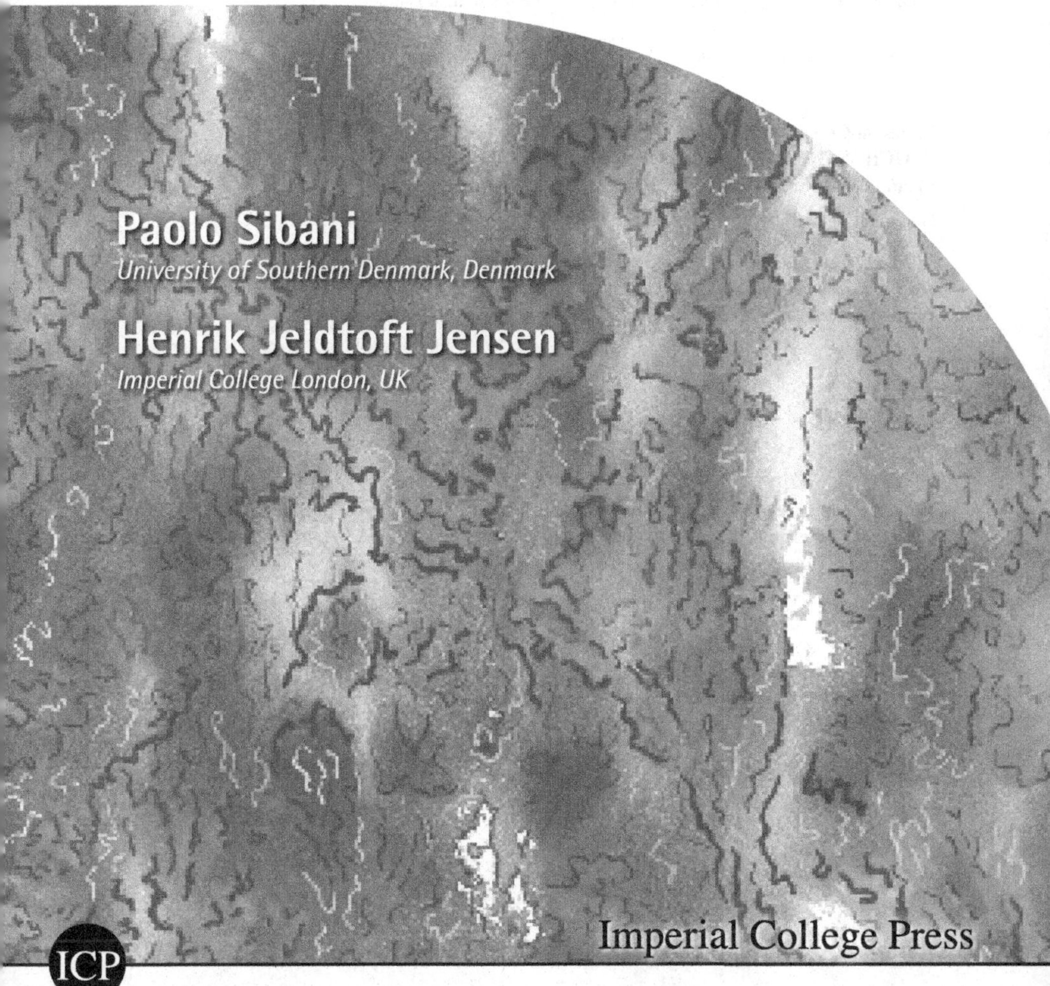

Paolo Sibani
University of Southern Denmark, Denmark

Henrik Jeldtoft Jensen
Imperial College London, UK

ICP

Imperial College Press

Published by

Imperial College Press
57 Shelton Street
Covent Garden
London WC2H 9HE

Distributed by

World Scientific Publishing Co. Pte. Ltd.
5 Toh Tuck Link, Singapore 596224
USA office: 27 Warren Street, Suite 401-402, Hackensack, NJ 07601
UK office: 57 Shelton Street, Covent Garden, London WC2H 9HE

British Library Cataloguing-in-Publication Data
A catalogue record for this book is available from the British Library.

Series on Complexity Science — Vol. 2
STOCHASTIC DYNAMICS OF COMPLEX SYSTEMS
From Glasses to Evolution

ISBN 978-1-84816-993-7

Typeset by Stallion Press
Email: enquiries@stallionpress.com

Printed in Singapore by World Scientific Printers.

Preface

The term 'complex system' appearing in the title of this book undoubtedly means different things to different people. If one conventionally describes a system in terms of interacting entities, i.e. particles or, more generally, degrees of freedom, many, including the authors of this book, would agree that its degree of complexity hinges on the number of degrees of freedom being very large and on the interactions lacking simple symmetries, such as the translational symmetry of a crystal. In this view, an amorphous material, e.g. a glass, is more complex than a crystal. Furthermore, complex systems cannot be completely isolated from their environment and are therefore subject to random influences beyond the observer's control. For this reason, a complex system is typically described by statistical properties which can be estimated from measurements or from computer simulations of suitable stochastic models. Chaotic systems are fully deterministic dynamical systems with only few degrees of freedom. They are hence not complex in the above sense, nor is a quantum mechanical system such as the hydrogen atom. Truly enough, quantum mechanical predictions are probabilistic in nature, but the uncertainty expressed by say, the Heisenberg relations is not rooted in our ignorance of the accidentals affecting the history of the system but reflects instead intrinsic limits to which all descriptions of nature at very small length scales are subject.

Many books exist about complex systems. Most are concerned with structure and other stationary properties, and when dynamical effects are considered, they do not lead to essential change of macroscopic properties. Consider for example the magnetic fluctuations of a material, kept exactly at its magnetic phase transition point. The average magnetization is equal to zero but the probability distribution of the size of the magnetic fluctuations has very interesting properties which reflect an incipient change of symmetry of the system, but do not change in time. The main focus of this book is on dynamics and especially on dynamics leading to change, i.e. on evolution and non-stationary states. We are concerned with slow but persistent changes observable at the macroscopic or systemic level. Unlike the magnet just mentioned, systems with these properties lack time translational symmetry.

One may say the book's focus is narrow, or one may perhaps agree that dynamical evolution over long time scales is a prominent feature of all the systems we intuitively think of as complex, say ecosystems, the brain or the economy. In the physics literature the term *ageing* refers to slow changes of observable properties which occur over time scales much longer than the patience, or indeed the lifetime of the observer. In the book, we use and develop this term for complex systems outside the realm of physics.

The book is divided into two parts. The material in the first part attempts to provide the necessary mathematical and computational tools and the intuition needed to deal with the systems described in the second part. Initially, the topics are of the standard sort, and are hence covered in several other books, but the emphasis and selection reflect both the author's interests and the overall theme of the book. Our discussion of Record Dynamics is new and is expected to be of relevance to a broad range of very different complex systems. The first part is suitable as teaching material for a one semester course dealing with the application of stochastic processes to natural science.

The second part of the book contains an introduction to the scientific literature and deals in some detail with the description of complex phenomena of physical and biological nature, e.g, disordered magnetic materials, superconductors and glasses, and models of co-evolution in ecosystems and even of ant behaviour. These rather heterogeneous topics are all dealt with using similar techniques of analysis. For reasons of space, scope, bias and of intended readership, this part does not aspire to be a scientific review of the subjects covered. We believe that one or more of its chapters can be used as basic material in a graduate course on complex systems and that researchers in any field of science who want to enter the complexity arena might find it interesting as well.

Throughout the book, each chapter begins with a qualitative introduction to the subject matter. It is our hope that these introductions will be useful not only for students and specialists following a course, but also for general readers with some basic knowledge of mathematics. Taken together, they can provide a background for the current hectic discussions about how complexity science can help to address some impelling problems in the world of today.

May 2012, Odense and London
Paolo Sibani and Henrik Jeldtoft Jensen

Acknowledgements

The present book is the outcome of a pleasant and long-standing collaboration between the authors. Beyond that, it reflects viewpoints on physics and complex systems which have evolved over the years through interactions with very many people. When it comes to the actual contents of the book, collaborations and discussions with colleagues and students have been of great importance. Paolo Sibani would like to thank Preben Alstrøm, Christian M. Andersen, Stefan Boettcher, Michael Brandt, Jesper Dall, John A. Hertz, Karl Heinz Hoffmann, Gregory G. Kenning, Peter Littlewood, Mark E.J. Newman, Richard G. Palmer, Andreas Pedersen, the late Jacob M. Pedersen, Jørgen B. Pedersen, Michel R. Schmidt, Christian Schön and Peter Salamon for their input on physics, science and all the rest. The many students who have been subjected to various drafts of the book and contributed with constructive criticism are commended for their patience and stamina in dealing with an unfinished work. A special thanks goes to Nikolaj Becker for carefully proof-reading parts of the manuscript and for delivering one of its figures. Henrik Jeldtoft Jensen is indebted to Tomas Alarcon, Kim Christensen, Simone A. di Collobiano, Matt Hall, Simon Laird, Daniel Lawson, Mario Nicodemi, Gunnar Pruessner, Tom O. Richardson and Ana Sendova-Franks. Both authors would like to thank Paul Anderson, Dominic Jones and Louis P. de Oliveira for their valuable input.

Families are important in all aspects of life, including the writing of a book. Paolo Sibani is indebted to his wife Karin who did his share of housework on top of her own and to our three kids, Nicolas, Camilla and Claudia, for cheering him up by doing well in all that they do. Henrik Jeldtoft Jensen is as always intrigued and overwhelmed by the persistent love and support from the world's three most wonderful women: his wife and daughters Vibeke N. Hansen, Barbara N.J. Jensen and Rebecca N.J. Jensen.

May 2012, Odense and London,
Paolo Sibani and Henrik Jeldtoft Jensen

Contents

Index of Figures

1

Introduction

The general view behind this book is that complexity science is a meaningful field of endeavour in which stochastic processes play an essential and unifying role. We therefore develop from scratch the methods and techniques needed to model a body of experimental and numerical data behind complex phenomena. These methods and techniques are then applied to a range of physical and biological problems of current interest.

HIERARCHIES, COMPLEXITY AND DYNAMICS

That science has an overall hierarchical structure, often related to characteristic length, time and energy scales, is fairly uncontroversial. Within this structure, many natural phenomena are only associated to a single or a few scales and can hence be studied in relative isolation. This fact underlies the success of the reductionist approach common in physics, where branches belonging to different levels of description have largely been developed independently, e.g. the elastic properties of an iron bar can be described without knowledge of nuclear physics, because nuclear processes and sound propagation occur on widely separated energy and length scales. While the particular organization of elementary particles making up the bar is the result of a series of dynamical processes, starting from nuclear reactions in a star and ending with its forging from an iron melt, all these processes are of little relevance to elasticity.

In contrast, energy and length scales are less broadly distributed and more closely intertwined in biology, e.g. a multicellular organism has itself a hierarchical structure: The dynamics of macromolecules leads to the formation of organelles and other cellular building blocks. The dynamical interactions of cells produces organs and all the different types of tissue. The entire structure of tissue and organs—helped by external agents such as bacteria—brings about a biological organism, e.g. a mouse or a man. Dynamics on much longer time scales is also important for structure: In the words of Theodosius Dobzhansky [Dob73], *"Nothing in biology makes sense except in the light of evolution."*

Darwinian evolution itself is a prime example of complex dynamics. First of all, it is inherently stochastic, i.e. it is driven by random mutations. Successful mutations owe their success to interaction patterns between individuals and between individuals and their environment. These patterns are mainly the outcome of historical contingencies and hence themselves random, to a degree. Secondly, even though in its original formulation Darwinian evolution only deals with individual organisms, the distinction between an individual and a tightly knitted ecosystem appears now blurred, and evolutionary principles are often applied at the level of species or even higher taxa. Hence, evolutionary dynamics is a random process which unfolds in a hierarchically organized structure, starting with cells at the lowest level and ending with, e.g., genera [Gou02].

While an engineer interested in the stiffness of a bicycle frame can safely neglect that the frame can melt, the situation differs in biology (and sociology and economics and neuroscience etc.), where phenomena at different scales do interact more strongly. Very small vira (size $10^{-8} - 10^{-6}$ m) can affect much larger living organisms, e.g. a blue whale, (linear size 10 m) or an Aspen cluster, (linear size 10^3 m). This possibility arises because, irrespective of body size, all cells have similar biochemical and biophysical properties. Furthermore, organisms are more than the sum of their parts: The functioning of the brain is not immediately reducible to that of its neurons. *In vitro* analyses may therefore miss important aspects of the *in vivo* behaviour, where several levels of a description hierarchy are inextricably mixed.

COMPLEX DYNAMICS

While a reductionist approach strives toward understanding and controlling what lies below the phenomena observed at some level in a hierarchical description of nature, the syncretic approach more common in complexity science strives to identify collective or emergent properties. Both procedures lead to new ontological levels: Since the times of Leucippus and Democritus the meaning of indivisible (atom) has been revised downwards many times and new particles have been identified along the way. Conversely, the fundamental frequency of a violin string is not visible at the level of nuclei, nor is the temperature defined at the level of a single degree of freedom. Both quantities are emerging properties at a description level involving a very large number of interacting degrees of freedom.

To move either down or up in the hierarchy of natural phenomena requires different methods. Moving down entails isolating the phenomenon to be described, by fixing a large number of 'external' parameters, e.g. optical tweezers can be used to study the elastic properties of a DNA molecule tethered to a bead. In contrast, the description at the emergent level involves the identification of relevant *coarse-grained* variables, whereby huge amounts of information are neglected, e.g. to describe the elastic vibrations of a violin string, the field of elastic deformations is needed but not the positions of the individual atoms.

Coarse-graining permeates much of physics, including areas not conventionally associated to complexity, e.g. ice and vapour contain the same water molecules but have widely different mechanical properties. Hence, the differences do not stem from the individual components, but from the nature of their interactions. Describing the phase diagram of water is a task for equilibrium statistical mechanics. The latter relies on postulating that microscopic configurations with the same energy are equiprobable, a probabilistic principle which coarse-grains away individual molecular trajectories. Finding emerging properties does in general require statistical methods and probability theory, elements which both embody some kind of coarse-graining in a physical modelling context.

Complex dynamics as defined in this book arises when processes belonging to many different levels of a hierarchy are intermingled and all contribute, in the long run, to a class of macroscopic phenomena. We argue that complex dynamical phenomena have important similarities straddling physics and biology.

That dynamics is an essential part of complex phenomena is the guiding principle of Self-Organized Criticality (SOC) [BTW87, Jen98, Pru11], a paradigm inspired by equilibrium critical phenomena, and aspiring to an all-encompassing generality. SOC emphasises that large classes of driven systems generically enter a stationary state whose real space structure is scale invariant. In contrast to critical phenomena, this happens with no need to tune any external parameters, whence the 'self-organized' part of the name. SOC focusses on the properties of the broad size distribution of certain intermittent events, called avalanches. These spatial rearrangements bring the system from one of its critical states to another. Similarly to SOC, we propose that a specific class of rare and intermittent events, which we call *quakes*, control the dynamics of many complex systems. Unlike SOC, the systems of interest to us are in a non-stationary and ever evolving state. We are interested in a statistical characterization of both the temporal distribution of the quakes and of the system changes they trigger, and claim that these properties are largely insensitive to the particulars of the system at hand. We are not overly concerned with self-similarity under real space dilation, although self-similarity turns out to be of importance. In general, the interactions between the parts of a complex system generate dynamical processes characterized by a hierarchy of time scales [Sim62]. In some cases the latter can be linked to a hierarchy in real space, in others to a hierarchy in a high dimensional configuration space, and in others again to an organizational hierarchy.

A hierarchy of length scales is of importance in critical phenomena: At the critical temperature, i.e. the temperature above which the distinction between liquid and gas is no longer meaningful, density fluctuations are present at all length scales. We can think of a large droplet of liquid water containing vapour, containing smaller droplets of liquid and so on, *ad infinitum*, or, more precisely, down to the molecular length scale.

A hierarchy of energy and time scales characterizes 'glassy' materials. As in SOC, the situation arises for a range of control parameters, and does not presuppose the fine-tuning of, e.g., the temperature. Glassy materials never reach true thermal equilibrium, but are instead in a permanent state of slow flux, a process known in the literature as physical ageing. Ageing is associated to a number of interesting properties, e.g. intermittency and so-called memory and rejuvenation effects. Intermittent events, the quakes just mentioned, accompany qualitative changes in the measurable properties of an ageing system. Successive quakes are separated by gradually longer periods of time during which the system appears quiescent at the level of macroscopic variables.

Biological evolution can be studied at the largest scales, millions of years, by analysing palaeontological data. On much shorter scales, years and even months, one can study bacterial and viral evolution. Finally, one can construct theoretical models and study them on the computer. No matter what the scale is, it is fairly clear that biological structures are metastable, i.e. evolve in time. At the level of say, individuals, evolution is a fairly continuous process involving small changes only. However, at the macroscopic level, say species, changes are rapid (on geological scales) and intermittent, and may involve large-scale reorganization. Some events, e.g. mass extinctions, are clearly due to causes external to the system's own dynamics, such as meteorite impacts. It is then of interest to study the reaction to the randomising effect of events which typically disrupt, to a lesser or greater extent, the pre-existing organization of an ecosystem. The tempo of evolutionary processes might be decreasing on palaeontological scales, and at least for computer models, there is a striking statistical similarity between the quakes leading a physical system from one metastable configuration to another and the analogous punctuations in an evolving biological system.

EMERGENT PROPERTIES IN COMPLEX SYSTEMS

One way to investigate the phenomena emerging at system level is to develop all-encompassing models and simulate them on a computer. Starting from the level of the individual component one attempts to include as many details as possible. Refined techniques are then used to compute the temporal evolution of huge numbers of degrees of freedom. The aim is to establish a faithful representation of the system considered, say transport through a major city, which allows virtual experiments to be carried out, leading to a detailed phenomenology. In our specific example one might try to collect minute information about the behaviour and needs of the citizens, e.g. weather conditions, fluctuations in fuel prices and their influence on transport preferences, etc. These simulations map out what to expect under given circumstances, which is obviously, a strength. The weakness is that the emphasis on high precision makes it difficult to discriminate between essential and less essential modelling features. This may in turn obfuscate which findings are of general relevance and which are only relevant for the specific situation at hand.

Simulation strategies designed to identify the *most* important mechanism responsible for certain phenomena and thereby hopefully point to more general aspects often start from bold and simplistic assumptions. In physics, this approach has been very successful for manifold reasons. Firstly, as earlier discussed, one may only need to focus on a few effective degrees of freedom, e.g. when studying the acoustics of a concert hall, one needs not worry about the electronic state of the air molecules. In fact one doesn't even need to worry about the molecular nature of the air, but can make use of a continuum description, e.g. fluid dynamics. This leads to a representation in terms of densities and waves rather than a representation in terms of individual air molecules. As a bonus, the study of reflection and refraction of waves in the concert hall may also be relevant to other forms of wave propagation.

While both types of approach are needed, they will typically serve different purposes. Detailed simulations can have great practical value, e.g. weather prediction. The more simplistic representation emphasizes the essential aspects allegedly responsible for the observed behaviour, and may elucidate aspects of relevance to a larger class of problems.

The anticipation that similarities exist across complex systems which greatly differ at the microscopic level is supported by our experience from mathematics and physics. An example of a greatly general principle, or law, that is equally relevant to, say, physics, sociology and biology is the Central Limit Theorem: When suitably scaled, the sum of many independent random variables has a normal probability distribution. This kind of universal behaviour comes about because the systemic variables, say the body mass of a biological organism, sum up the contributions from all the parts comprising the total. The macroscopic or systemic behaviour is a consequence of mathematics and will not depend on specific details. Clearly such 'laws' are only applicable if the relationship between the components and the collective macroscopic level for a given system is in fact consistent with the assumptions underlying the mathematical result. In the case of the Central Limit Theorem applicability demands sufficiently weak interactions between the constituents.

Phase transitions display a spectacular example of similarity between very different physical systems. The behaviour can be understood in terms of the Renormalization Group (RG) analysis, a quantitative way to establish what the relevant system components are at different length scales, and how these components interact with one another. As earlier mentioned, at a specific value of certain parameters, e.g. temperature and pressure, the macroscopic behaviour at large length scales turns out to be the same irrespective of microscopic details. These lessons from mathematics and physics suggest that at the collective systemic level complex systems of different origin may well look similar. In our view, complexity science has a role to play in the identification and study of such emergent regularities in systems far from equilibrium.

We expect the study of emergent phenomena and the identification of generalities across different classes of problems to be more challenging in biology than in physics. Biological systems are always evolving and their description is intrinsically a stochastic process, as is the case in Darwinian evolution. We also expect that any 'laws' for complex system dynamics will be statistical in nature, and thus be similar in character to Ohm's law, which is a relation between macroscopic quantities, current and voltage, and which comes about as a result of averaging over the scattering of the individual electrons as they collide with inhomogeneities in a conducting material.

THE REST OF THIS BOOK

The first part of the book, which includes Chapters 1 to 8, is mainly dedicated to the development of general concepts and methods within statistical data analysis and stochastic processes. These chapters contain exercises, examples and problems. The first three can be used for a short introductory course on stochastic processes in the natural sciences. The following five introduce more advanced topics, which are either applications or developments of the material already presented. Taken together, the eight chapters are suitable for a one semester course.

In the second part of the book, each of Chapters 9 to 13 contains a brief monographic description of a different area in complex science. The topics are selected because of the authors' personal involvement and interest, and because taken together, they buttress the claim implied by the book's title on the unity of complex dynamics. The material in the second part is more advanced. It can serve as an introduction to the scientific literature and it can be used for graduate level courses in complex system dynamics.

All chapters start with a brief introductory section. These sections provide a synopsis of the book's contents which does not rely on mathematics and which can be read independently of the rest.

Finally, a technical note: Spelling is mainly British, but the original American spelling is kept in quotations, in the bibliography and in figure texts. Numerical simulation results are usually expressed in terms of dimensionless quantities. In other cases, units are specified as needed.

COPYRIGHTS AND CREDITS

The forty figures included in the book are meant to illustrate central points of the material. Some were especially prepared, but others are either reprinted or adapted from previously published material. We extend our grateful thanks to the copyright holders, authors and publishers, for kindly and promptly granting us the permission to reproduce their figures. A full bibliographic reference to the original publication is always given in the caption of a reproduced figure.

Figures reproduced from *Physical Review Letters, Physical Review B* and *Physical Review E* are all copyrighted by the American Physical Society in the year of their publication. Readers may view, browse, and/or download these figures for temporary copying purposes only, provided these uses are for noncommercial personal purposes. Except as provided by law, the figures may not be further reproduced, distributed, transmitted, modified, adapted, performed, displayed, published, or sold in whole or part, without prior written permission from the American Physical Society. Figures reproduced from *EPL* are copyrighted by EDP Sciences in the year of their publication. Figures reproduced from the *Journal of Statistical Physics* and the *Journal of Physics: Condensed Matter*, are copyrighted by IOP Publishing Limited, Bristol, UK, in the year of their publication. For figures reproduced from a publication co-authored by Paolo Sibani in the *Journal of Theoretical Biology*, the copyright holder Elsevier grants the "right to ... re-use" them "in other works, with full acknowledgement of its original publication in the journal". See http://www.elsevier.com/wps/find/authorsview.authors/rights.

Part I

Complex Dynamics: Tools and Applications

2

Characterization of Collective Dynamics

2.1. INTRODUCTION

Information on, e.g., stock prices, the weather and seismic events is often conveyed through *time series,* time ordered sequences of measurements or observations empirically characterizing some dynamical processes of interest. The statistical techniques described in this chapter help to extract from these data useful information on the system's dynamics. Crucial for modelling is to ascertain whether the dynamics is stationary or not, e.g. climate changes are related to the stationarity or lack thereof of time series of measured temperature values.

Stationarity means that observations fluctuate reversibly around a constant value. This suggests[1] that the system itself could be in a steady or equilibrium state. In contrast, one or more so-called one point averages, e.g. the average and the variance, are time dependent in a non-stationary time series. To analyse non-stationary series may require an approach quite different from what is discussed in the present chapter, see, e.g., Chapter 6.

This chapter starts out with a qualitative discussion of why statistical methods can be applied to collective properties of natural

[1] We purposely avoid the stronger 'implies' since measuring two different properties of the same systems may yield a stationary, respectively non-stationary series.

systems. A more technical part is then concerned with classical techniques for stationary time series, especially the use of Fourier transforms for estimating the correlation function. Some knowledge of basic concepts of probability theory is assumed throughout.

2.2. BASIC CONSIDERATIONS

If available information is insufficient to fully determine the time development of a system, probabilistic approaches are called for. Predictions are then statistical in nature, i.e. loosely speaking, they describe typical rather than actual outcomes.

It is seldom possible or desirable to specify the state of a natural system in full. At the quantum mechanical level, performing a measurement already affects the state of the system. Furthermore, quantum mechanical equations deal with *probability amplitudes* and their predictions are inherently probabilistic. A classical approach must come to terms with the ubiquitous presence of *chaos,* where the slightest uncertainty in the initial state grows exponentially in time. Accurate long time predictions are therefore impossible and a probabilistic element creeps in.

Systems with a large number of interacting components (particles, agents, etc.) can be described at different levels of detail. The most detailed, but often least useful, level is the *microscopic* level, where the time dependent state of each component is specified.

> *Example 1:* Enumerating the centre of mass coordinates of a mole of diatomic molecules at a single instant requires approximately $6 \times 6 \times 10^{24}$ real numbers, or about 3×10^{17} Gbyte of storage. Rotational and electronic degrees of freedom make the list even longer. This huge amount of information is impossible to obtain and to process, and a microscopic description is useless for predictive purposes.

A useful *thermodynamic* or *macroscopic* description only requires a handful of variables, e.g. for a physical system, temperature, pressure, and chemical potential. These do not reflect properties of individual molecules or system components but rather describe collective, i.e. statistical, properties of the system as a whole. They become useful as gradually larger systems are considered, and for this reason are often called *emergent* properties. In particular, thermodynamic variables are closely associated to statistical *averages*, e.g. the temperature of a gas is related to the average kinetic energy of its constituent molecules. Between the microscopic and macroscopic levels lies a *mesoscopic* description, which includes fluctuations and is concerned with correlation functions and other statistical properties.

A statistical approach to multicomponent systems is amply justified by its great practical applicability. Conceptually, two (related) arguments can be used to rationalize its success. The *first* argument is that, due to insufficient time resolution, experimental data represent time averages rather than truly instantaneous measurements. According to the *ergodic* hypothesis, to which we shall return, these time averages are equivalent to averages taken at a single instant of time over a collection of independently evolving systems, a so-called *ensemble*. The elements of the ensemble share a certain set of macroscopic properties, e.g. all the elements of the *microcanonical* ensemble share the same volume, energy and number of constituents. They differ however with respect to their microstates. The microstates are thus *constrained*, rather than determined, by the macroscopic properties of the ensemble to which they belong. The fundamental probabilistic hypothesis of statistical physics states that the microscopic configurations compatible with a given set of constraints are all equiprobable.

Since our first argument produces a time independent probability distribution, it does not apply to out of equilibrium situations. A *second* argument, which partly relies on the first, and which more readily extends to time dependent processes, is based on the necessity to identify the system as an entity separated from its 'environment'. While the system proper, if isolated, might be amenable to a deterministic microscopic description, e.g. because of the limited number of variables involved, the environment is usually not amenable to such a description. The interactions between the two parts can only be described in probabilistic language, and the dynamics hence acquires a probabilistic character.

In the following example, the interactions among the system components are imperfectly known, and a statistical description is from the outset the only possibility.

> *Example 2:* A large democratic country has $M = 100$ million voters and two political parties, for concreteness the Red (R) party and the Blue (B) party. Since each voter has two possible political choices, the population of voters has $2^{100000000} \approx 10^{24}$ possible states. These are the microstates of the electorate. A voter's choice is influenced by, e.g., family traditions, educational background and the opinions of friends, plus of course the political programmes of the two parties and the personalities of their leaders. A deterministic model for how the electors would cast their ballots is hardly conceivable. A feasible mesoscopic description considers instead the probabilities that a voter would remain faithful to his or her 'old' party, or change party at the next election. This gives four different probabilities R→R, R→B, B→R and B→B. These four probabilities can be (unreliably) estimated using opinion polls, whence probabilities can be calculated for the outcome of the next election. Finally, a macroscopic description is concerned with the total number of electors voting for R and B, and with how these numbers would be effected by, e.g., the unemployment rate, the GNP, international conflicts, and other factors such as the weather on election day.

In the following, x identifies a system configuration, t is the time, and $x(t)$ is a sequence of states generated by the dynamics, a so-called trajectory. Each configuration is specified by, e.g., a vector or scalar quantity. A deterministic dynamical rule uniquely associates a trajectory $x(t|x_0)$ to any starting point x_0. On the other hand, a probabilistic, or stochastic rule, assigns probabilities to the transitions between a state and its possible successors. In this case, a recorded trajectory is only one of many possible outcomes or *realizations* of the *stochastic process* $x(t)$ describing the system dynamics.

The most general description available for a stochastic process is provided by the conditional probability $P(x, t|\{x(t')\}_{t'<t})$ of state x at time t given the dynamical history for $t' < t$. The situation is simpler for so-called Markovian processes, where the probability of the current state suffices to determine the *probability distribution* for future microstates. A Markovian process is fully described by a transition probability matrix, with elements given by the conditional probability $P(x, t|x', t')$ for state x at time t, given state x' at time t'. Key to both deterministic and Markovian descriptions is that sufficient information be included in the definition of a state.

A mesoscopic model involves a reduced set of states, and the transition probabilities between these states. Crucial elements in the modelling process are the collection and statistical analysis of the time resolved information contained in a *time series* of measurements.

2.3. TIME SERIES AND THEIR STATISTICAL PROPERTIES

A time series is a discrete sequence of numbers $x_1, x_2, \ldots x_k$, typically obtained by recording data at regularly spaced intervals of time, i.e. every δt. Then, $x_k = x(t_k)$, with $t_k = \delta t \cdot k$. A time series $\{x_k\}$ samples a particular realization, or trajectory, of the stochastic process $x(t)$.

For a statistical characterization of a time series with N terms, consider M independent repetitions of the N measurements, $x_k^{(i)}, i = 1, 2 \ldots M; k = 1, 2 \ldots N$, where each series of measurements is taken under identical conditions. The ensemble introduced in the previous section is the collection of M trajectories in the limit $M \to \infty$. The value measured at time t_k from the i'th trajectory is denoted by $x_k^{(i)}$. The mathematical *expectation value* for $x(t)$, $E(x(t))$, is defined by the limit

$$\mu(t) \stackrel{\text{def}}{=} \mathrm{E}(x(t)) \stackrel{\text{def}}{=} \lim_{M\to\infty} \frac{1}{M} \sum_{i=1}^{M} x^{(i)}(t). \tag{2.1}$$

The variance of $x(t)$ is defined as

$$\sigma^2(t) \stackrel{\text{def}}{=} \mathrm{Var}(x(t)) \stackrel{\text{def}}{=} E(x(t)^2) - E^2(x(t)). \tag{2.2}$$

The correlation function involves the expectation value of the product of two values $x(t)$ and $x(t')$

$$\mathrm{Corr}(t, t') = E(x(t)x(t')) - E(x(t))E(x(t')). \tag{2.3}$$

The first two quantities are functions of the time t and the third is a function of the two times t and t'. For discrete series, the same definition applies, with $x(t)$ replaced by $x_k = x(t_k)$, e.g. $\mathrm{E}(x_k) = \mathrm{E}(x(t_k))$.

2.3.1. Stationary Time Series

If neither the expectation value $\mathrm{E}(x(t))$ nor any other one point moments $\mathrm{E}((x^l(t)), l = 2, 3, \ldots$ depend on time, the stochastic process and the relative time series are called *stationary*. A stationary process has time independent average μ and variance σ^2. The average can conveniently be subtracted, yielding a process with zero average. In the following, zero average for stationary series is assumed unless otherwise specified. Furthermore, we invoke ergodicity and extract statistical properties of interest from a single time series.

The correlation function

Correlation functions convey information on the interdependence between different entries of a time series. Let us first consider a stationary time series $x(t_k)$ whose elements are statistically independent and identically distributed stochastic variables. Examples are lottery extractions, or the durations of the intervals between successive decays of a radioactive nucleus, with their averages duly subtracted. That entries x_k and x_j are independent implies that $\mathrm{E}(x_j x_k) = \mathrm{E}(x_j)\mathrm{E}(x_k)$ for $k \neq j$. Hence, using that the average of each element is zero, one obtains

$$\mathrm{E}(x_j x_k) = \mathrm{E}(x^2)\delta_{j,k} = \sigma^2 \delta_{j,k}. \tag{2.4}$$

The so-called Kronecker delta $\delta_{j,k}$ equals one if $j = k$ and is otherwise zero. In the following example, the value of an element of the series is strongly related to the preceding values.

Example 3: A convicted felon is detained at home. He is allowed to drive without stopping between home and work, and his position and speed are monitored every minute by a GPS device. Walking speeds are below the detection threshold of the GPS, and the corresponding recordings are all zeros. The 60 km commute to work takes approximately one hour each way. Assume that the speed data are recorded in the time series $\{x_k\}$. Clearly, the series comprises two long subseries with zero entries, covering the hours spent at work and at home. These are intercalated by shorter subseries where the recorded speed values fluctuates around the commuting average speed of 60 km per hour. An additional random variation arises as the starting time of the commutes can vary by up to one hour.

Assume now that one recording, e.g. the first, is $x_1 = 0$. The person is then either at home or in his office at the time of the measurement. Accordingly, the values $x_2, x_3 \ldots$ in the series are more likely to be zero than not. Conversely, if $x_1 = 30$ km per hour, the person is driving, and the following values are likely to be in the same range. Since the value of a given entry carries information about the possible values of the following and preceding entries, the time series of this example is correlated.

In the above example, the data has a cyclic structure, which extends through the workweek (and beyond). Figure (2.1) *(a)* shows a series of recordings of yearly average surface temperature in the Arctic

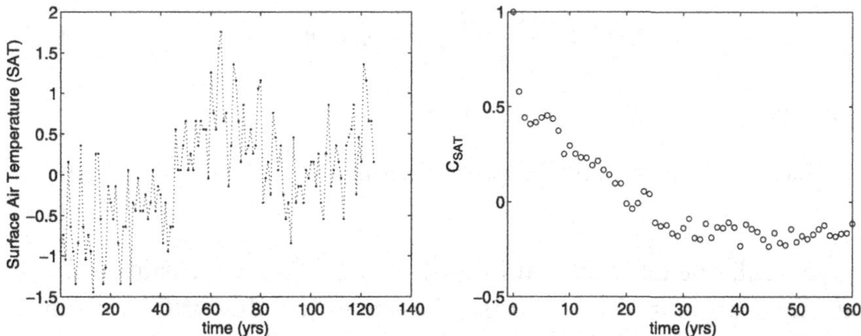

Figure 2.1. The left figure shows the fluctuations of the Surface Air Temperature (SAT) in the Arctic region from 1875 to 1999. All data points are yearly averages with the arithmetic mean of the temperature over the 124 year time span subtracted. The right figure shows the correlation function of these data, calculated at the inverse Fourier transform of the power spectrum, as discussed in the main text. There is no attempt to de-trend the data. Data taken from Ref. [Soo05]

region, with the arithmetic mean over a time span of 124 years subtracted. The origin corresponds to the year 1875. Panel *(b)* of the same figure shows the correlation function, calculated using the Wiener–Khintchine theorem (discussed later). No attempt has been made to analyse these data for trends, and the series has been treated as a stationary series. Data taken from Ref. [Soo05].

The interdependence of the elements of a series is described by the discrete version of Eq. (2.3):

$$\mathrm{Corr}(j,k) = \mathrm{E}(x_j x_k) - \mu^2. \tag{2.5}$$

For a stationary series with zero average, the second term on the r.h.s. of Eq. (2.5) vanishes. Secondly, both indexes can be shifted arbitrarily by the same amount without changing the function, i.e. $\mathrm{Corr}(j,k) = \mathrm{Corr}(0, k-j)$. Keeping only the difference as argument, the correlation function can be re-written as

$$\mathrm{Corr}(n) = \mathrm{E}(x_0 x_n) = \mathrm{Corr}(-n); \quad -N < n < N. \tag{2.6}$$

Since $\mathrm{Corr}(n = 0) = \sigma^2$, correlation functions are often normalized by dividing through by the variance. The normalized and dimensionless discrete correlation function is then

$$C(n) = \frac{\mathrm{Corr}(n)}{\sigma^2}, \quad -N < n < N. \tag{2.7}$$

Clearly, $C(0) = 1$ and $C(-n) = C(n)$.

Exercise 1: Using the property $\mathrm{E}([x(t_k) - x(t_0)]^2) \geq 0$ show that $C(n) \leq 1$.

In general, one expects that $\mathrm{E}(x_k x_l) \to \mu^2$, when the distance $|k-l|$ becomes large. In this situation, the absolute value of C eventually decays to zero. If different x_k values represent states of a physical system, the memory of a certain state does not extend arbitrarily far into the future. A lingering memory of past states is in fact an important characteristic of a complex dynamical system. For the moment, however, this possibility will be ignored and the correlation function will be assumed to have a fast, e.g. exponential, decay.

Asymptotically for large N values, the quantity

$$l_C = 1 + 2 \sum_{n=1}^{N-1} C(n) \tag{2.8}$$

becomes independent of N. For an exponentially decaying correlation function, l_C can be interpreted as a *correlation length*. On the scale of l_C, the correlation decays to zero, i.e. when n is a small integer, e.g. $n = 5$, $C(nl_C) \approx 0$. For a series of independent random numbers, only the first term contributes to the r.h.s. of Eq. (2.8), and $l_C = 1$.

Estimation of statistical properties

Expectation values and correlation functions can only be calculated analytically if the distributions characterizing the series are known. A numerical evaluation can be accurate if a large, in principle infinite, collection of independent experimental data sets is at one's disposal. Usually, neither condition is fulfilled, and the expectation values and other statistical properties of interest must be estimated using a limited amount of available data, e.g. a single time series. A large amount of statistical literature is devoted to this, sometimes surprisingly complicated, task.

Consider again the, admittedly trivial, case of a stationary series of *independent* and identically distributed random numbers. In this case, the series is a sample of a stochastic variable x. To estimate the expectation value μ from this sample, the arithmetic mean

$$\overline{x}(N) = \frac{1}{N} \sum_{k=1}^{N} x_k \tag{2.9}$$

can be used. For any N, the latter provides a so-called *unbiased estimator* of $\mathrm{E}(x)$.

Consider a certain statistical property P of the sample, e.g. an expectation value, which we wish to estimate from sample data. An unbiased estimator $\hat{P}$ is a stochastic variable whose values can be calculated from samples, and which satisfies $\mathrm{E}(\hat{P}) = P$. The latter property tells us that the difference between the estimated and the correct value of P is symmetrically distributed around zero. The property says nothing about the magnitude of the error, and, hence, about the quality of the estimator. The variance $\mathrm{Var}(\hat{P})$ can be used to gauge the quality of the estimator, since its value is zero if and only if the estimation is perfectly accurate.

Exercise 2:

1. A series $\{x_k\}$ consists of N independent and identically distributed random variables with expectation value μ and variance σ^2. Using only basic properties of the expectation value, show that the arithmetic mean $\overline{x}(N)$ is an unbiased estimator of μ.
2. Argue that the same result remains valid even if the elements of the series are not independent.
3. Returning to the case of independent random numbers, show that

$$\mathrm{Var}(\overline{x}(N)) = \frac{1}{N}\mathrm{Var}(x) = \frac{1}{N}\sigma^2. \tag{2.10}$$

   This result does not, in general, hold if the elements of the series are not independent.
4. What happens with the quality of the estimator when the series $\{x_k\}$ becomes very long?

Let us now return to a series of correlated values. The variance of the arithmetic mean can be expressed as

$$\mathrm{Var}(\overline{x}) = \frac{l_C}{N}\sigma^2, \tag{2.11}$$

where l_C is given in Eq. (2.8). Accordingly a good estimate of the expectation value of a stationary series requires that the series be much longer than the correlation length, i.e. $N \gg l_C$. This is only possible if the correlation function $C(n)$ decays sufficiently fast for l_C to remain bounded in the limit $N \to \infty$.

Exercise 3:

1. Prove Eq. (2.11).
2. Assuming that $C(n)$ is fast decaying, i.e. that $l_C \ll N$, argue that the quantity $\hat{\sigma}^2$, given by

$$\hat{\sigma}^2 = \frac{1}{N - l_C} \sum_{k=1}^{N} (x_k - \overline{x})^2, \tag{2.12}$$

is an unbiased estimator of the variance σ^2 of the series $\{x_k\}$. Noting that $l_C = 1$ for independent random variables, the above result reduces in this case to a well-known formula.

Equation (2.12) is at first sight of limited use, since knowing l_C already requires a knowledge of the correlation function. We note however that for 'sufficiently' large N, i.e. $N \gg l_C$, the exact value of l_C is not important. It is also clear that the correlation function is crucially important in the analysis of a stationary time series. A convenient way to analyse the properties of the correlation function of a stochastic process is provided by the Wiener–Khintchine theorem, which exploits the properties of Fourier transforms.

The Wiener–Khintchine theorem

The Wiener–Khintchine (WK) theorem relates the correlation function of a stationary stochastic process to the square modulus of the Fourier transform of the signal, see Eq. (2.20). This theorem has enormous practical importance, as we shall see in the next chapter.

The treatment of the WK theorem is notationally simpler when using the time dependent process $x(t)$ rather than the series x_k. Secondly, by shifting the time axis, we define our process in an interval $[-T, T]$ symmetric around zero. Thirdly, we define the curtailed process x_T by $x_T(t) = x(t)$ for $|t| < T$, and zero otherwise. Finally, the 'observation' time $2T$ is assumed to be so large that estimated and exact statistical quantities can be identified. In the following, whenever the symbol T appears, the limit $T \to \infty$ is implied. In this section we replace, for simplicity, the symbol Corr for the correlation function, with the symbol C. The (unnormalized) correlation function is then given by

$$C(\tau) = \frac{1}{2T}\int_{-T}^{T-\tau} x(t)x(t+\tau)dt = \frac{1}{2T}\int_{-\infty}^{\infty} x_T(t)x_T(t+\tau)dt. \tag{2.13}$$

Extending the integration to the whole real axis allows us to use the properties of the Fourier transform $\mathcal{F}_f$ of a square integrable function f. The Fourier transform and its inverse are given by

$$\mathcal{F}_f(\omega) = \int_{-\infty}^{\infty} f(t)e^{-i\omega t}dt; \quad f(t) = \frac{1}{2\pi}\int_{-\infty}^{\infty} \mathcal{F}_f(\omega)e^{i\omega t}d\omega. \tag{2.14}$$

Equality between a function and the inverse of its Fourier transform holds in the l_2 norm. We also recall that the Fourier transform is an isometry, i.e. for any function in l_2

$$\int_{-\infty}^{\infty} |\mathcal{F}_f(\omega)|^2 d\omega = \int_{-\infty}^{\infty} |f(t)|^2 dt. \tag{2.15}$$

By Fourier transforming Eq. (2.13) we obtain the so-called power spectrum of the process

$$S(\omega) \stackrel{\text{def}}{=} \mathcal{F}_C(\omega) = \frac{1}{2T}|\mathcal{F}_{x_T}(\omega)|^2. \tag{2.16}$$

Equation (2.15) applied to the power spectrum, combined with the fact that $x(t)$ is a real function, tells us that

$$\int_{-\infty}^{\infty} S(\omega)d\omega = \frac{1}{2T}\int_{-\infty}^{\infty} x_T^2(t)dt = \overline{x^2}. \tag{2.17}$$

Assume for concreteness that $x(t)$ represents a voltage fluctuating across an Ohmic resistor. The last term on the r.h.s. is then proportional to the electric power dissipated in the resistor, and $S(\omega)$ is the power dissipated across the resistor per unit of frequency.

It follows directly from Eq. (2.13) that

$$S(\omega) = \int_{-\infty}^{\infty} C(\tau)e^{-i\omega\tau}d\tau. \tag{2.18}$$

In particular, reintroducing the distinction between the unnormalized and normalized version of the correlation function,

$$S(0) = \int_{-\infty}^{\infty} \mathrm{Corr}(\tau)d\tau = \sigma^2 \int_{-\infty}^{\infty} C(\tau)d\tau \tag{2.19}$$

is widely used to estimate the correlation time, cf Eq. (2.8), a quantity denoting the scale on which the correlation decays to zero. Note, however, that the correlation function might not be integrable, in which case the correlation time is undefined. A divergent integral signals the presence of a scale invariant dynamical process. In this situation, $\lim_{\omega\to 0} S(\omega) = \infty$.

The correlation function can now be calculated from the power spectrum by applying the inverse Fourier transform

$$C(\tau) = \frac{1}{2\pi}\int_{-\infty}^{\infty} S(\omega)e^{i\omega\tau}d\omega = \frac{1}{\pi}\int_{0}^{\infty} S(\omega)\cos(\omega\tau)d\omega. \tag{2.20}$$

The last equality relies on S being an even function of ω.

The shape of the power spectrum

Since the correlation function is even, Eq. (2.18) can be written as

$$S(\omega) = 2\int_{0}^{\infty} C(t)\cos(\omega t)dt. \tag{2.21}$$

In many cases, the correlation function decays (at least asymptotically for large times) as an exponential function, characterized by a time scale τ, i.e. $C(t) = \sigma^2 \exp(- \mid t/\tau \mid)$. The corresponding power spectrum has then a Lorentzian shape

$$S(\omega) = \sigma^2 \frac{2\tau}{1 + (\omega\tau)^2}. \tag{2.22}$$

Note that the spectrum is flat for $\omega\tau \ll 1$. For large enough values of ω, $S(\omega) \propto \omega^{-2}$, a form usually called $1/f^2$ noise, with reference to the cyclic frequency $f = \omega/(2\pi)$. Thus, this particular shape of the power spectrum indicates that the relaxation process is dominated by a single time scale, corresponding to an exponential decay of the correlation function. Consider now the case in which the correlation function decays as a power-law, i.e. $C(t) = (1 + t/\tau)^{-\alpha}$, where $0 < \alpha < 1$.

A power-law decay is scale invariant, and lacks, in other words, any characteristic scale. The power spectrum has then the form

$$S(\omega) \propto \omega^{-1+\alpha}. \tag{2.23}$$

A power-law spectrum with a decay exponent numerically close to one is called $1/f$ noise. This form can be indicative of long-ranged correlations in the stationary process producing the time series at hand. Note however that when very long relaxation times are present in the dynamics, a stationary regime is hard to achieve. To be sure that the system is stationary, i.e. that the slowest relaxation mode has decayed to zero, the spectrum must be flat in the low end of the frequency range.

Discrete series and discrete Fourier series

In practice, only a finite discrete time series is usually available, i.e. the observation interval $2T$ is finite, and so is the sampling interval δt. The Fourier transform is correspondingly discrete. Discrete Fourier transforms are easily calculated with commercially available software. The finiteness of the series also means that the issue of statistical estimation cannot be ignored.

Consider a discrete series $\{x_k\}, 0 \le k \le N$, with $\mathrm{E}(x_k) = 0$ and $\mathrm{Var}(x_k) = \sigma^2$. The frequency ranges from zero to $\frac{2\pi}{\delta t}(1-1/N)$ in steps of $\frac{2\pi}{N\delta t}$. Correspondingly, time ranges from zero to $T = (N-1)\delta t$ in steps of δt. To ease the notation we set in the following $\delta t = 1$. Given a set of N discrete frequencies $\omega_l = 2\pi l/N$ for $l = 0, \ldots N-1$, the elements of the (complex) discrete Fourier series $\{\tilde{x}(\omega_l)\}$ have the form

$$\tilde{x}(\omega_l) = \sum_{k=0}^{N-1} x_k e^{-ik\omega_l}. \tag{2.24}$$

The inverse Fourier transform is then

$$x_k = \frac{1}{N} \sum_{l=0}^{N-1} \tilde{x}(\omega_l) e^{ik\omega_l}. \tag{2.25}$$

The Fourier representation of the data enforces a periodicity (with period N) on the extension of the series beyond the original interval and on its Fourier coefficients. Since the x_k values are stochastic variables, so are the Fourier coefficients $\tilde{x}(\omega_l)$. The following properties then hold:

$$\mathrm{E}(\tilde{x}(\omega_k)) = 0 \tag{2.26}$$

$$\mathrm{E}(|\tilde{x}(\omega_k)|^2) = N \sum_{q=-(N-1)}^{N-1} C(q)e^{-i\omega_k q}. \tag{2.27}$$

From the second equation above, it follows that

$$C(k) = \frac{1}{N}\sum_{l=0}^{N-1} \frac{\mathrm{E}(|\tilde{x}(\omega_l)|^2)}{N} e^{i\omega_l k}. \tag{2.28}$$

As done for the continuous case, we now define the power spectrum of x as

$$S(\omega_l) = \frac{|\tilde{x}(\omega_l)|^2}{N}. \tag{2.29}$$

The function

$$\hat{C}(k) = \frac{1}{N}\sum_{l=0}^{N-1} S(\omega_l)e^{i\omega_l k} \tag{2.30}$$

is clearly periodic with period N. $\hat{C}(k)$ is an unbiased estimator of the correlation function due to Eq. (2.27). Since the original data series $\{x_k\}$ is real, its Fourier coefficients are even functions of ω. The power spectrum is therefore a real and even function of ω, and $\hat{C}(k)$ is an even function of k, as expected. Assuming for convenience that N is even, the evenness and periodicity of $\hat{C}$ imply $C(N/2+k) = C(N/2-k)$, i.e. the correlation function estimator is symmetric around the midpoint of the observation interval. Estimating the variance of $\hat{C}$ involves four point correlation functions, and is left as an exercise for the mathematically inclined reader. Since $\{x_k\}$ is real, the corresponding Fourier coefficients are even functions of their argument, i.e. $\tilde{x}(\omega_l) = \tilde{x}(-\omega_l)$. The same is true for the power spectrum $S(\omega_l)$.

Exercise 4:

1. Each term of a time series $\{z_k\}, k = 0 \ldots N-1$ can be written as

$$z_k = \sin(\omega k) + n_k \tag{2.31}$$

where each element n_k is independent of all the others and uniformly distributed in the interval $(-1/2, 1/2)$. Calculate the power spectrum of the series $\{z_k\}$, analytically and numerically for a suitable value of N.

Exercise 5:

1. Assume for convenience that the length N of a series is an even number. Argue that, due to the symmetry properties of the correlation function $C(n)$, only $N/2$ of its values convey dynamical information. The other $N/2$ values are simply given by symmetry.
2. In principle, whether or not the series $\{x_n\}$ has zero average or not only affects the Fourier coefficient at $l = 0$. Numerically, it is nevertheless often a good idea to ensure that the series has zero average. Why?
3. Argue that for small frequencies near $\omega = \pi/(N\delta t)$, (we have here reinstated the physical dimension of the frequency) the values of the power spectrum $S(\omega_l)$ are spurious, since they reflect the periodicity of $C(n)$ imposed by the finiteness of N.

Exercise 6:

1. Using Eqs. (2.27) and (2.29), show that the power spectrum $S(\omega)$ of a stationary of series of independent random numbers $\{x_k\}$ is independent of ω, i.e. $S(\omega) = S$. A series with a flat power spectrum is often called *white noise*.
2. Conclude that, if x_k is a white noise signal of length N, then, to leading order in N, $\tilde{x}(\omega_l) = S^{1/2}N^{1/2}e^{i\theta(l,N)}$, where $\theta(l, N)$ is a random number.

Exercise 7: According to Eq. (2.19), the value of the power spectrum taken at $\omega = 0$ equals the correlation time of the corresponding time series $x(t)$. What is then **wrong** with the following argument:

> The stationary time series $x(t)$ has zero expectation value. (If not, we can shift it by a constant to obtain a zero expectation value.) Therefore, $\mathcal{F}_x(\omega = 0) = 0$ and, due to Eq. (2.16), $S(0) = 0$. In conclusion, any stationary series has zero correlation length.

Hint: Argue that, for a series with zero expectation value the stochastic variable

$$S_T = \frac{1}{T}\left|\int_{-T}^{T} x(t)dt\right|^2 \neq 0$$

has, in the limit $T \to \infty$, a distribution which is independent of T.

Fourier series and power spectra are applicable tools, irrespective of the origin and nature of the time series. Yet, in connection with stochastic processes, they are only safe to use in the stationary case. Since this limitation is often ignored, it is of some interest to pinpoint where, precisely, stationarity comes into play. Technically, Eq. (2.27) is no longer valid if the process is not stationary, because the correlation function then has two arguments, rather than one. Ignoring this warning sign, we proceed and, using Eq. (2.24), form the series $\{\tilde{x}(\omega_l)\}_{l=0\ldots N-1}$. The time series $\{x(t_k)\}_{k=0\ldots N-1}$ reconstructed from the Fourier series via Eq. (2.25) is always periodic with period N. In the stationary case, this is a reasonable approximation: Replicating the finite series already obtained *ad infinitum*, rather than continuing to measure new data points, produces a new stationary series. The errors introduced are zero on average, and furthermore, they become small when N is sufficiently large. In the non-stationary case, however, the original non-stationary series is replaced with a stationary one. This generates a systematic error, i.e. an incorrect average, and should be avoided.

2.3.2. Non-stationary Time Series

In a non-stationary time series an equilibrium (or stationary) state may altogether be lacking, or may exist but not be attained yet within the observation time covered by the series. Time translational invariance is absent in both cases, at least one moment depends on time, and correlation functions depend on two time arguments. Furthermore, information about the 'initial' state is never fully lost, and might even be recovered by a suitable analysis. Non-stationary series describe non-stationary processes, of which many examples are found in nature, e.g. most importantly, biological evolution.

Techniques which work for stationary series sometimes produce misleading results if inadvertently applied to non-stationary series. We shall return to how non-stationary series and non-stationary stochastic processes can be treated in different situations. Unfortunately, general techniques are few and far between. Accordingly, this section is mainly concerned with simple introductory examples highlighting the differences between the stationary and non-stationary case.

Our first example deals with the position of a Brownian particle. The average of the corresponding time series is zero, while the variance increases linearly with time.

Example 4: A simple non-stationary time series $\{x_n\}$ is the position of a so-called *Brownian particle*. For simplicity we consider the projection of the particle's motion along the x-axis. Consider times $t_1, t_2 \ldots t_k$ with $t_k - t_{k-1} = \Delta t$. Let Δ_k be the incremental displacement between times t_{k-1} and t_k. The position of the particle at time t_n is then the sum of a *white noise* sequence

$$x_n = \sum_{k=1}^{n} \Delta_k. \tag{2.32}$$

The increments Δ_k of the Brownian particle are independent and identically distributed stochastic variables with zero expectation and with variance $\mathrm{Var}(\Delta_k) = 2D\Delta t$, where D is the diffusion coefficient. Accordingly, $\mathrm{E}(x_n) = 0$ and $\mathrm{Var}(x_n) = 2Dt_n$. The variance of x_n increases linearly in time, and linearly with the index n. Hence, the series is not stationary.

To show that a blind application to non-stationary series of techniques developed for stationary cases produces dubious results, consider the following exercise.

Exercise 8:

1. The correlation function of the position is

$$\mathrm{Corr}(k,l) = \mathrm{E}(x_k, x_l) = \sum_{i=1}^{k}\sum_{j=1}^{l} \mathrm{E}(\Delta_i \Delta_j). \tag{2.33}$$

   Show that $\mathrm{Corr}(k,l) = 2D\Delta t \min(k,l)$.
2. The arithmetic mean of the position of the Brownian particle is, as always, defined as $\overline{x} = N^{-1}\sum_{n=1}^{N} x_n$. Show that $\overline{x}$ is an unbiased estimator of the expectation value of the particle's position.
3. The variance of this estimator is given by

$$\mathrm{Var}(\overline{x}) = \frac{1}{N^2}\sum_{k=1}^{N}\sum_{l=1}^{N} \mathrm{Corr}(k,l). \tag{2.34}$$

   Show that the variance does not approach zero for $N \to \infty$ and contrast this behaviour with the case of a stationary series described by Eq. (2.11). Is the arithmetic mean accurate as an estimator of the expected position of the particle?

We note in passing that once the particle is confined in a region of linear size L the characteristic 'diffusive' behaviour of the Brownian particle disappears on a time scale $\tau_L = \frac{L^2}{2D}$. In this case, waiting long enough will restore a stationary behaviour.

Our second example deals with a system fluctuating 'at random' around a time dependent average $\mu(t)$.

Example 5: The time series $\{z_k\}$ has the form

$$z_k = \Delta_k + \mu_k, \quad k = 0 \ldots N-1, \tag{2.35}$$

where $\{\Delta_k\}$ is white noise with unit variance and where $\mu_k \stackrel{\mathrm{def}}{=} \mu(t_k)$ is the value of a known function μ at time $t_k = k\delta t$. The expectation value and correlation function of this series are

$$\mathrm{E}_z(k) = \mu_k \tag{2.36}$$

$$\mathrm{Corr}_z(k,l) = \mathrm{Corr}_\Delta(k,l) = \delta_{k,l}. \tag{2.37}$$

The power spectrum of $\{z_k\}$ is therefore flat:

$$S_z(\omega_l) = S_\Delta(\omega_l) = 1. \tag{2.38}$$

Let us now pretend that the series $\{z_k\}$ is stationary and apply the machinery of the previous section. We easily find

$$S_z(\omega_l) = 1 + \frac{1}{N}\left(2\mathcal{R}\left(\tilde{\Delta}(\omega_l)\tilde{\mu}(\omega_l)\right) + |\tilde{\mu}(\omega_l)|^2\right). \tag{2.39}$$

At low frequencies, the 'power spectrum' above can easily be dominated by a divergent Fourier transform of μ and the outcome of the procedure then bears no relationship to the correct result, which is given by Eq. (2.38).

Let us now consider the above example in more detail. To that end we choose a slowly changing average

$$\mu(t) = 20 \cdot t^{\alpha},$$

where $-1 < \alpha < 0$. The three plots in Fig. (2.2) show the power spectrum, naively calculated, of the non-stationary white noise series given in Eq. (2.35). The data plotted are averages over 400 independent realizations of a series of Δ_k. Each realization contains 2^{18} elements, has $\Delta t = 1$ and the same additive term $\mu(t)$. The flat high frequency part of the spectrum extends upwards to all frequencies

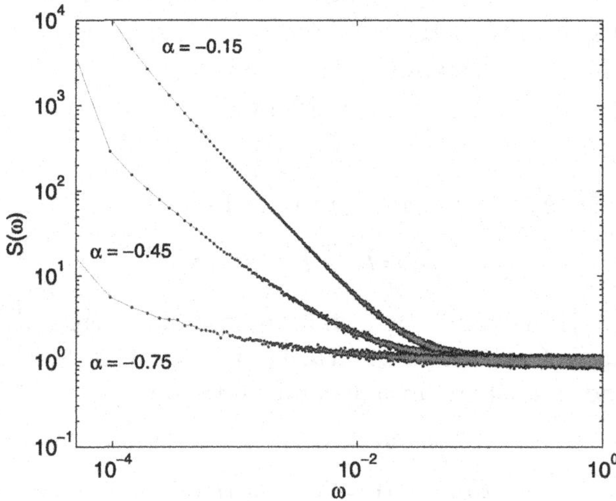

Figure 2.2. The three power spectra are calculated for time series with a slowly changing average as if the series were stationary. This leads to spurious results.

up to 2π. Especially for $\alpha = -0.15$, the very slow drift of the average is well masked in the time domain by the random fluctuations of the data. In the frequency domain it produces the low frequency part of the spectrum, which diverges as a power-law. This part could well be erroneously interpreted as $1/f$ noise of a stationary series.

To understand aspects of the observed low frequency behaviour qualitatively on the basis of Eq. (2.39), we first note that the lowest frequency in a spectrum measured over a time span of length t is of the order of $\omega_{\text{min}} = \mathcal{O}(1/t) \propto 1/N$. We therefore consider the limits $N \to \infty, \omega \to 0$, with $\omega N = \mathcal{O}(1)$, and neglect all proportionality constants. Since $\tilde{\mu}(\omega)$ scales for low frequencies as $\omega^{-\alpha-1}$, the last term in Eq. (2.39) scales as $\omega^{-2\alpha-1}$. This term diverges for $\alpha > -1/2$ in the limit $\omega \to 0$. The midterm has a similar asymptotic behaviour, due to $\tilde{\Delta}(\omega_l) = \mathcal{O}(N^{1/2})$.

The admittedly rough arguments given above indicate that, as observed, the closer α is to zero, the more prominent the low frequency divergence of the power spectrum will be. This is rather worrying, as the extremely slow drift of the signal described by a power-law with an exponent close to zero is hardly detectable in the time domain. To be sure that a time series is actually stationary on the time scale of observation, one should see a low frequency cut-off: Below the cut-off frequency, the power spectrum should be flat, meaning that even the slowest relaxation mode has decayed to zero. Not surprisingly, the power spectra shown in Fig. (2.2) does not feature any low frequency cut-off. Were the spectra part of the analysis of actual data, this should be reason enough to regard them with a healthy dose of scepticism.

2.3.3. Stationary or Non-stationary?

To conclude, we return to the important issue already hinted at in the introductory section: How can we be sure whether a time series is stationary or not. There is no fireproof answer, but the power spectrum of the series again comes to our help.

As explained in Chapter 3, the correlation function of a Markov process can, under rather general conditions, be expanded as a superposition of exponential functions, each with a different decay time. The equation for the correlation function of some stationary process F, Eq. (3.87), is duplicated below for convenience, and reads

$$\mathrm{Corr}_F(t) = \sum_{\lambda<0} \left(\sum_i F(i) P_{\mathrm{eq}}^{1/2}(i) \psi_\lambda(i) \right)^2 e^{\lambda t}. \tag{2.40}$$

The sum is over different negative eigenvalues, and excludes the equilibrium eigenvalue $\lambda = 0$. Each term is a decaying exponential, multiplied by a positive weight. The largest eigenvalue present in the sum corresponds to the relaxation time $\tau = 1/|\lambda_{\mathrm{max}}|$. Let us first assume that λ_{max} is bounded away from zero, i.e. that a positive real number ϵ exists, such that $\lambda_{\mathrm{max}} < -\epsilon$. For long enough times, $t > 1/\epsilon$, the correlation function is dominated by the exponential term with the longest decay time. We can therefore repeat the steps leading to Eq. (2.22), and recognize that, for small enough frequencies, i.e. $\omega < \epsilon$, the power spectrum is flat: $S(\omega) = \text{constant}$. Hence, if a power spectrum is constant below a certain low frequency cut-off, we know that a stationary state is reached, and that the form of the power spectrum above the cut-off genuinely reflects the properties of the correlation function. Suppose, however, that no cut-off is seen, as, e.g., is the case in Fig. (2.2). Two possibilities arise: *i)* There would eventually be a cut-off, but it lies, alas, below the lowest frequency we can observe, i.e. below the reciprocal of the observation time. If a trustworthy friend now vouches for the series being stationary,[2] we can still relate its power spectrum to properties of the correlation function. *ii)* The eigenvalues of the Markov process underlying the data have an accumulation point at zero. In this case, regardless of the length of the observation time, the power spectrum has no cut-off. Even though we might still be able to argue that the average of the process is constant in time, e.g. equal to zero on symmetry grounds, applying the Wiener–Khintchine theorem in this situation is inherently unsafe, because the series is inherently non-stationary.

[2]If lacking a friend, repeating the measurements with a longer observation time is also a possibility.

3

Markovian Stochastic Processes

3.1. INTRODUCTION

A deterministic description of complex dynamical systems is prevented by the very large number of degrees of freedom present, usually combined with an imperfect knowledge of their interactions. Furthermore, it is desirable to model the dynamics in terms of a limited number of variables. All this naturally leads to stochastic processes as modelling tools at a level intermediate between an (unfeasible and undesirable) microscopic description and a purely macroscopic and deterministic description, e.g. fluid dynamics and time dependent thermodynamics. At the so-called mesoscopic level, one keeps information on fluctuations, and predictions are statistical in nature: A famous example is the Brownian particle, e.g. a grain of pollen in a glass of water, jostled around by much smaller water molecules. Although we cannot predict where the particle will be at a certain point of time, we can calculate the probability that it will be at a given position, given a certain starting point. In a fully deterministic dynamical description, knowledge of the initial state uniquely determines the time evolution. The dynamics can be run backwards, and the initial state can be reconstructed from any point along the trajectory emanating from it. Hence, perfect memory of the initial state is retained. Not so in a stochastic process, where such memory is eventually destroyed, in part or in full, by a series of random changes.

In systems near stable configurations, random fluctuations of dynamical variables or parameters usually only act as perturbations of deterministic macroscopic equations, which by themselves account for the main qualitative features of the dynamics. In other cases, however, fluctuations control the long time dynamical behaviour, e.g. random genetic variation and natural selection are the basic mechanism of biological evolution, a paradigmatic example of complex dynamics. A less grand example of the far-reaching consequences of fluctuations is the so-called 'gambler's ruin' problem: A gambler plays a fair game against an infinitely wealthy opponent. His odds are 50%, and his net average gain (or loss) appears at first to be zero. However, with probability one, the gambler will lose all his capital, and the game will terminate. Clearly, fluctuations are crucially important in this case.

Beside being useful descriptions of natural phenomena with an inherent element of randomness, stochastic processes can be used as computational tools in different areas, such as optimization problems of scientific and industrial relevance. Simulation techniques embodying Markov chains, a certain type of stochastic process, are major inroads into complex systems dynamics.

Basic material on stochastic processes is widely available in the literature. We refer to the books of Feller [Fel66] and Van Kampen [Kam92] for a more thorough mathematical discussion and for additional applications to physics and chemistry. Excellent books are also available on the application of Markov chains to computational problems of various kinds, including optimization [NB99, SSF02]. This chapter provides a concise introduction to the subject matter and the background needed for our later discussions of specific problems in complex dynamics.

We start with the Langevin equation, which offers a simple and intuitive approach to a limited class of Markov processes. Secondly, we consider general properties of Markov processes, including linear response theory, which is of paramount importance in probing complex dynamics, see, e.g., the discussion in Chapter 9. Thirdly, the master equation is introduced and exemplified, and its convergence properties and those of Markov chains are discussed. A brief treatment of Fokker–Planck equations concludes the chapter.

3.2. LANGEVIN EQUATIONS

Langevin equations apply to deterministic dynamical systems additively perturbed by a rapidly fluctuating random term of zero average. A classic example is the velocity $\mathbf{v}(t)$ of a small particle immersed in a fluid. Restricting ourselves for simplicity to the x component of the velocity, the Langevin equation for this problem reads

$$m\frac{dv}{dt} = -\gamma v + F(t). \tag{3.1}$$

The equation embodies Newton's second law, since its l.h.s. is the particle's mass m multiplied by the acceleration, and its r.h.s. represents the total force acting on the particle. The latter consists of a friction term $-\gamma v$ and of a term $F(t)$ representing a stochastic process of zero average. The assumed rapid variation of F reflects the numerous collisions of the Langevin particle with the molecules in the surrounding fluid. The correlation function of F is assumed to decay very quickly on the time scale $\tau = m/\gamma$. The dissipative term $-\gamma v$ also stems from the collisions with the fluid, and F and γ are hence related through a so-called Einstein relation, as discussed below.

3.2.1. Solution Techniques for the Langevin Equation

The solution technique for the Langevin equation crucially relies on the separation between microscopic and macroscopic time scales. Its success buttresses the idea that the fluid surrounding the Langevin particle is composed of small molecules which provide the random force. Today this idea is commonplace, and yet it was vigorously contested by prominent physicists in the late years of the 19th century.

Using the notation

$$\tau \stackrel{\text{def}}{=} m/\gamma \quad \text{and} \quad f \stackrel{\text{def}}{=} F/m, \tag{3.2}$$

the formal solution of Eq. (3.1) with initial condition v_0 is

$$v(t) = v_0 e^{-\frac{t}{\tau}} + \int_0^t e^{\frac{t'-t}{\tau}} f(t')dt'. \tag{3.3}$$

On time scales $t \gg \tau$, the exponentially decaying term on the r.h.s. can be safely neglected. The short decorrelation time of the Langevin term $f(t)$ is captured by approximating f with *white noise,* i.e. by assuming that the autocorrelation function has the form

$$\langle f(t')f(t'')\rangle = C^2\delta(t' - t''), \tag{3.4}$$

where C is a constant.

Exercise 1: Using Eq. (3.3), argue that for times long compared to τ, $v(t)$ is itself, to a good approximation, a stationary process with zero average.

Exercise 2: Show that the variance $\sigma_v^2(t)$ of the particle velocity approaches a constant for large t.

Fourier transforming Eq. (3.1) leads to

$$\mathcal{F}_v(\omega) = \frac{\tau \mathcal{F}_f(\omega)}{1 + i\omega\tau}. \tag{3.5}$$

The power spectrum $S_v(\omega)$ of the particle velocity is connected to $S_f(\omega)$, the power spectrum of the noise, by

$$S_v(\omega) = \frac{\tau^2 S_f(\omega)}{1 + (\omega\tau)^2} \tag{3.6}$$

where, due to Eq. (2.18), $S_f(\omega) = C^2$.

Exercise 3: Check that a velocity autocorrelation function of the form

$$\text{Corr}_v(t) = \sigma_v^2 e^{-\frac{|t|}{\tau}}, \tag{3.7}$$

leads to Eq. (3.6), provided that

$$\sigma_v^2 = \frac{\tau C^2}{2}. \tag{3.8}$$

The stationary solution of the Langevin equation describes the velocity fluctuations of a thermalized particle. As the particle velocity has zero expectation, $\sigma_v^2 = \overline{v^2}$ is twice the thermal average of the kinetic energy of the particle, divided by the particle's mass. Using the *equipartition theorem* from equilibrium statistical physics, in combination with Eq. (3.8), we infer that

$$C^2 = \frac{2k_B T}{\tau m} = \frac{2k_B T \gamma}{m^2}, \tag{3.9}$$

where T is the temperature and k_B is the Boltzmann constant. This is the anticipated relation between the statistical properties of the noise, here expressed through the power spectrum of f, and the damping constant γ.

3.2.2. Ballistic and Diffusive Limit

The position of the Langevin particle is given by the integral of Eq. (3.3) with respect to time

$$x(t) - x_0 = \int_0^t v(t')dt'.$$

For short times, $t \ll \tau$, $v(t) = v_0 + \mathcal{O}(t)$ and

$$x(t) = v_0 t + \mathcal{O}(t^2). \tag{3.10}$$

Exercise 4: Show that Eq. (3.10) is valid. Why is this regime called *ballistic*?

For time scales $t \gg \tau$, the mathematical expectation of both velocity and position vanishes. Assuming that the velocity correlation function C_v is nearly zero except in a small interval of width $\Delta \approx \tau$, which is symmetric around zero, the variance of the position is

$$\sigma_x^2(t) = \int_0^t \int_0^t E(v(t')v(t''))dt'dt'' \approx t \int_{-\Delta}^{\Delta} C_v(z)dz. \tag{3.11}$$

In the above formula, the integration is carried out in a diagonal strip of length $t\sqrt{2}$ and width 2Δ. This domain differs from the support of the integrand near the points $(0,0)$ and (t,t). However, the error introduced is of order Δ^2, and can be neglected for $t \gg \Delta$. As the integrand in Eq. (3.11) vanishes outside the interval $[-\Delta, \Delta]$ no harm is done in extending the integration to the whole real axis. Using Eq. (2.19), we finally obtain

$$\sigma_x^2(t) = S_v(0)t \approx \tau^2 S_f(0)t = \frac{2k_B T\tau}{m}t. \tag{3.12}$$

The variance of the position grows linearly in time. This defining characteristic of *Brownian motion*, is usually expressed as $\sigma_x^2(t) = 2Dt$, where the constant D is called the diffusion coefficient or diffusion constant. Equating the two expressions just obtained for the variance, we obtain

$$D = \frac{k_B T\tau}{m} = \frac{k_B T}{\gamma}. \tag{3.13}$$

Exercise 5: Detail the mathematical steps leading to Eq. (3.11), using, e.g., a drawing of the region of integration.

Exercise 6: Equation (3.13) implies a relation between the diffusion coefficient and the radius of the (supposedly spherical) Langevin particle. Consult the literature to find this relation.

Exercise 7: A particle immersed in a fluid is elastically bound to a fixed point, for convenience taken as the origin. The restoring force on the particle has the form $-kx$, where x is the particle displacement. The fluid provides a damping force $-\gamma\frac{dx}{dt}$ as well as a delta correlated Langevin force $F(t)$, with zero average and with $\text{Corr}_F(t) = m^2C^2\delta(t)$. Write down the Langevin equation for this problem. Solve the equation in the frequency domain using the Fourier transform. Finally, use the equipartition theorem to link the autocorrelation function of the Langevin force to the temperature of the fluid.

3.3. GENERAL PROPERTIES OF MARKOV PROCESSES

The *state* or *configuration* space of a stochastic process comprises all its possible configurations. The space is equipped with a topology, or neighbourhood relation, where neighbouring states are connected by elementary moves, or transitions. The process is accordingly characterized by a set of *transition rates* or transition probabilities connecting neighbouring states on microscopic time scales. This information is often deducible from more fundamental descriptions, e.g. quantum mechanics and sometimes simply postulated in the modelling assumptions.

Example 1: The Ising model in its many variants has applications ranging from physics to social science. Its state space consists of tuples

$$(\sigma_1, \sigma_2, \ldots \sigma_M),$$

where σ_i is a dichotomic variable, usually called 'spin' in physical applications. Traditionally the variables assume values ± 1 or $\{0, 1\}$. Configuration space hence consists of the 2^M vertices of an M-dimensional hypercube. Neighboring states differ in the value of precisely one σ_i. This configuration space can be viewed as a graph whose nodes are the states, and where edges connect neighbouring states. For a dynamical description, transition probabilities must be assigned to each edge.

Importantly, a real space structure can also be associated to each configuration, e.g. when the σ_i are located on a d-dimensional lattice. Configuration space and real space properties are both important, but should not be confused.

On mesoscopic and macroscopic time scales, one is interested in the conditional probabilities of finding the system in state x_n at time t_n, given the history of the states previously visited at times $t_{n-1}, t_{n-2}, \ldots t_1$. We write these conditional probabilities as $T_n(x_n, t_n|x_{n-1}, t_{n-1}; x_{n-2}, t_{n-2}; \ldots x_1, t_1)$. The distribution $T_1(x_1, t)$ describes the probability distribution over the states at some 'initial' time t. The probabilities obey a set of *deterministic* equations, which can in principle be solved, providing a statistical description of the dynamics.

In a Markov process the transition probability depends on the 'current' state and not on the past trajectory in its entirety,[1] i.e.

$$T_n(x_n, t_n|x_{n-1}, t_{n-1}; x_{n-2}, t_{n-2}; \ldots x_1, t_1) = T_2(x_n, t_n|x_{n-1}, t_{n-1}).$$

The conditional probability $T_2(x_2, t_2|x_1, t_1)$ of a Markov process is here referred to as the *propagator*. The term *kernel* is used in the statistical literature. The Markovian character of the process seemingly introduces a huge simplification. However, unless the configuration space of the problem is specified, the simplification may well be illusory, as illustrated in the following example.

Example 2: Consider M point particles moving in physical space according to the laws of classical mechanics. The microscopic states which make up *phase space* are specified by tuples

$$\mathbf{Q} \stackrel{\text{def}}{=} (\mathbf{r}_1, \mathbf{r}_2, \ldots \mathbf{r}_M; \mathbf{p}_1, \mathbf{p}_2, \ldots \mathbf{p}_M)$$

where $\mathbf{r}_i$ is the position of the i'th particle and where $\mathbf{p}_i$ is the corresponding linear momentum. Let $\mathbf{f}(t, \mathbf{Q}_0)$ be the phase space trajectory passing through the point $\mathbf{Q}_0$. The propagator is in this case

$$T_2(\mathbf{Q}_2, t_2|\mathbf{Q}_1, t_1) = \delta(\mathbf{Q}_2 - \mathbf{f}(t_2 - t_1, \mathbf{Q}_1)). \tag{3.14}$$

The above slightly contrived example illustrates that a Markovian description is always formally achievable at the expense of including more information in the definition of a state. The real challenge is to find an approximate Markovian description using a minimum of microscopic information.

---

[1]It then follows that the subscript n on $T_n(x_n, t_n|x_{n-1}, t_{n-1}; x_{n-2}, t_{n-2}; \ldots x_1, t_1)$ is entirely redundant. It will therefore mainly be omitted and/or used for different purposes, e.g. we write $T_2(x_2, t|x_1, 0)$ as $T(x_2, t|x_1)$.

The time dependence of the propagator is determined by the *master equation*. Below we consider properties of Markov processes which are independent of the precise form of the propagator, and also independent of configuration space properties such as the connectivity of configuration space. The latter properties will therefore often be left unspecified.

3.3.1. Master Equations and Markov Chains

In a Markov process with discrete states, the time evolution of $T(n_2, t_2|n_1, t_1)$ is controlled by the *master equation*. Since the form of T at time $t_2 = t_1$ is expressed by the initial condition of the master equation, the variable n_1 will often be omitted. The notation

$$T_{n,n_0}(t) \stackrel{\text{def}}{=} T(n, t|n_0, 0) \tag{3.15}$$

for the probability of state n, given the initial state n_0 stresses the matrix nature of T.

Let k_{ij} be the rate at which probability flows from state j to state i due to the elementary transitions in the dynamics. For a given set of k_{ij}, configuration space is structured as a directed graph, in which each node is a state, and where nodes j and i are connected by an edge if and only if $k_{ij} \neq 0$. In most cases, $k_{ij} \neq 0$ implies $k_{ji} \neq 0$. The configuration space is in this case represented by an undirected graph.

The master equation reads

$$\frac{dT_{nn_0}(t)}{dt} = \sum_i k_{ni} T_{in_0}(t) - \sum_i k_{in} T_{nn_0}(t), \tag{3.16}$$

with initial condition

$$T_{nn_0}(t = 0) = T(n, 0|n_0, 0) = \delta(n, n_0). \tag{3.17}$$

Equation (3.16) simply states that the net change of probability per unit of time arises as the balance between the probability in- and out-flow into the state. Suitable boundary and/or integrability conditions must also be supplied, as discussed in the examples.

In computer simulations, where the time variable is usually discrete, i.e. $t = 0, 1 \ldots$, a discrete version of the master equation is called for:

$$T_{nn_0}(t+1) = T_{nn_0}(t) + \sum_i p_{ni} T_{in_0}(t) - T_{nn_0}(t) \sum_i p_{in}. \tag{3.18}$$

The coefficient p_{ni} represents the probability of jumping from state i to n in a single step. The above equation defines a *Markov chain*.

The master equation can be written compactly using matrix notation. In the standard basis $\mathbf{e}_k, k = 1, 2 \ldots$, with $\mathbf{e}_k(j) = \delta_{j,k}$, we let $\mathbf{T}(t)$ be a matrix with elements $T_{in}(t)$, and $\mathbf{W}$ a matrix with off-diagonal elements

$$\mathbf{W}_{in} = k_{in}, \quad n \neq i. \tag{3.19}$$

Due to conservation of probability, the column sums of $\mathbf{W}$ must all vanish. Hence, its diagonal elements are given by the column sums

$$\mathbf{W}_{nn} = -\sum_{i \neq n} k_{in}. \tag{3.20}$$

The master equation now reads

$$\frac{d\mathbf{T}(t)}{dt} = \mathbf{WT} \tag{3.21}$$

where $\mathbf{W}$ is called *transition matrix*.

The content of Eq. (3.17) is expressed more compactly in matrix notation by the initial condition $\mathbf{T}(t = 0) = \mathbf{I}$, where $\mathbf{I}$ is the unit matrix. With this initial condition, the solution to Eq. (3.21) is simply

$$\mathbf{T}(t) = e^{t\mathbf{W}}. \tag{3.22}$$

From this expression, one easily derives the relation

$$\mathbf{T}(t + t') = \mathbf{T}(t)\mathbf{T}(t'). \tag{3.23}$$

Written out element wise, the equation reads

$$T_{nk}(t + t') = \sum_j T_{nj}(t) T_{jk}(t'), \tag{3.24}$$

a relation equivalent to Eq. (3.29), which is widely known as the Chapman–Kolomogorov relation.

The stationary nature of the equilibrium distribution P_{eq} implies that for all $t \geq 0$

$$\mathbf{T}(t)P_{\mathrm{eq}} = \left(1 + t\mathbf{W} + \frac{t^2\mathbf{W}^2}{2!} + \cdots\right)P_{\mathrm{eq}} = P_{\mathrm{eq}}, \tag{3.25}$$

from which

$$\mathbf{W}P_{\mathrm{eq}} = 0 \tag{3.26}$$

immediately follows. The equilibrium distribution is thus an eigenvector of the transition matrix $\mathbf{W}$, with eigenvalue zero.

Exercise 8: Show that the vanishing of all column sums is tantamount to conservation of probability.

Exercise 9: Construct an equation similar to Eq. (3.21) for the Markov chain, i.e. using integer valued rather than continuous time. Show that the column sums of the corresponding $\mathbf{W}$ matrix are in this case all equal to one. Interpret this result.

Exercise 10: Describe in words the probabilistic interpretation of the Chapman–Kolomogorov relation, Eq. (3.24).

To use Equation (3.22) in practice, the values of all rate coefficients must be provided. Even in a system with a moderate number of states, say 10^3 states, this is a daunting task. As discussed below, the principle of *detailed balance* somewhat constrains the number of free parameters. A further simplification can be attained by lumping together subsets of states, a process resulting in a coarse-grained problem with more sparsely connected states. The master equation can be approximated even further by so-called Fokker–Planck equations, i.e. partial differential equations with a modest number of free parameters, more easily determined by fitting to experimental data. Last but not least, some important properties of master equations depend only weakly on the specific form of the rate coefficients. This is the case for the fluctuation-dissipation theorem discussed below. Widely observed properties of non-stationary processes, e.g. *intermittency* and *memory behaviour* are exquisitely dynamical in nature, and can be modelled by master equations. Thus, even with imperfectly known parameters, master equations provide a theoretical framework to describe generic features of relaxation phenomena.

3.3.2. The Propagator and its Moments

Without further ado, labels x_i are treated below first as real numbers, and later as integers. Modest notational changes can easily accommodate more complicated situations, e.g. tuples of integers or reals.

In a stationary process the time origin can be chosen arbitrarily. Accordingly (and briefly returning to our initial notation for conditional probabilities), $T_1(x, t_1) = P_{\text{eq}}(x)$ is time independent, and $T_2(x_2, t_2|x_1, t_1) = T(x_2, t|x_1, 0)$ only depends on the time difference $t = t_2 - t_1$. The latter property, which is called *homogeneity*, is a necessary but not sufficient condition for stationarity, e.g. in a diffusion process in Euclidean space $T_2(x_2, t_2|x_1, t_1)$ only depends on $t = t_2 - t_1$, but no stationary or equilibrium state exists. Diffusion is therefore time homogeneous but not stationary.

Exercise 11: Discuss the following properties of the transition probability of a stationary Markov process:

$$\lim_{t\to 0} \quad T(x_2, t|x_1) = \delta(x_2 - x_1), \tag{3.27}$$

$$\lim_{t\to \infty} \quad T(x_2, t|x_1) = P_{\text{eq}}(x_2), \tag{3.28}$$

$$T(x_3, t + \tau|x_1) = \int T(x_3, t|x_2)T(x_2, \tau|x_1)dx_2. \tag{3.29}$$

The integral covers the full range of x_2. Taking the limit $t \to \infty$ in the last equation, show that

$$\int T(x_2, \tau|x_1)dx_2 = 1, \quad \forall x_1. \tag{3.30}$$

Taking the limit $\tau \to \infty$ in the same equation, show that

$$P_{\text{eq}}(x_3) = \int T(x_3, t|x_2) P_{\text{eq}}(x_2) dx_2. \tag{3.31}$$

Explain the result in words.

The moments of a Markov process can be calculated using $T(x_2, t|x_1)$ and $P_{\text{eq}}(x)$. To this end, consider an arbitrary function F defined on the configuration space. Without loss of generality, F will be assumed to have zero average in a *stationary* Markov process. This situation can always be achieved the by the shift $F \to F - \mu_{\text{F,eq}}$, where $\mu_{\text{F,eq}}$ is the equilibrium average of F. For an initial ensemble described by $P_{\text{eq}}(x)$ the expectation value of F is

$$\mu_{\text{F}}(t) = \iint F(x_3) T(x_3, t|x_2) P_{\text{eq}}(x_2) dx_2 dx_3. \tag{3.32}$$

Exercise 12: Using Equation (3.32) show that

$$\lim_{t\to\infty} \mu_{\text{F}}(t) = \mu_{\text{F,eq}} = \int F(x_3) P_{\text{eq}}(x_3) dx_3 = 0. \tag{3.33}$$

The autocorrelation function of F is defined as

$$\text{Corr}_F(t) = \int\int F(x_3) F(x_2) T(x_3, t|x_2) P_{\text{eq}}(x_2) dx_2 dx_3. \tag{3.34}$$

Exercise 13: Use Eq. (3.34) to show that

$$\lim_{t\to\infty} \text{Corr}_F(t) = 0 \tag{3.35}$$

and that

$$\text{Corr}_F(t=0) = \sigma^2_{\text{F,eq}}, \tag{3.36}$$

where $\sigma^2_{\text{F,eq}}$ is the variance of F in the stationary state. As the correlation function is often normalized to be unity in the initial state, we define

$$C_F(t) = \frac{\text{Corr}_F(t)}{\sigma^2_{\text{F,eq}}}. \tag{3.37}$$

Autocorrelation functions often provide valuable experimental information on stationary stochastic processes, as discussed later.

3.3.3. Fluctuation-dissipation Theorems

The *fluctuation-dissipation* theorems we are about to discuss rely on the existence of a stationary state P_{st}, to which all initial states supposedly converge, but do not rely on the form of the relaxation process leading to it.

In physical applications, the stationary distribution coincides with the equilibrium distribution P_{eq} known from equilibrium statistical mechanics. In a system thermally equilibrated at temperature T, P_{eq} is given by

$$P_{eq}(j) = \frac{e^{-\epsilon(j)}}{Z}, \tag{3.38}$$

with

$$\epsilon(j) = E(j)/(k_B T) = \beta E(j), \tag{3.39}$$

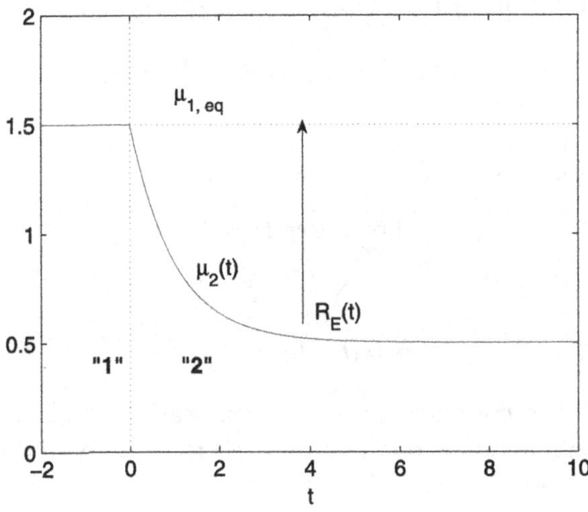

Figure 3.1. A change of temperature at time $t = 0$ induces the relaxation of the mean energy $\mu_2(t)$ from its initial equilibrium value $\mu_{1,eq}$. The response, indicated by the arrow, is given by Eq. (3.41).

where $\beta = 1/(k_B T)$ and where k_B, the Boltzmann constant, will usually be omitted for notational convenience. The denominator

$$Z = \sum_j e^{-\epsilon(j)}, \tag{3.40}$$

which provides the correct normalization of the equilibrium distribution, is called the *partition function*.

We now consider a system initially prepared in its stationary state, and assume that one or more parameters entering the transition matrix **W** are instantaneously modified by imposing at time zero a stepwise perturbation. As this creates an off equilibrium situation, a relaxation process begins eventually leading to the equilibrium state pertaining to the modified **W**. The average change of various observable quantities during this relaxation process embodies the system's *dynamical response* to the stepwise perturbation. The situation involves two equilibrium averages, which cannot both be set to zero, and which are therefore both kept in the notation.

We are now poised to analyse the linear response of a specific quantity, the energy, to a sudden change of the parameter β from β_1 to β_2. Henceforth, subscripts 1 and 2 refer to situation '1', where $t \le 0$, and the system is in equilibrium, and to situation '2', where $t > 0$ and a response evolves as just described, i.e. $\mu_{1,\mathrm{eq}}$ is the equilibrium average energy in state 1 and $\mu_2(t)$ is the average energy in state 2. The energy response to the change in β is

$$R_E(t) \stackrel{\mathrm{def}}{=} \mu_2(t) - \mu_{1,\mathrm{eq}}. \tag{3.41}$$

Clearly, $R_E(t = 0) = 0$ and $R_E(t = \infty) = \mu_{2,\mathrm{eq}} - \mu_{1,\mathrm{eq}}$. Consider now the auxiliary function

$$\Delta_E(t) \stackrel{\mathrm{def}}{=} \mu_2(t) - \mu_{2,\mathrm{eq}} = \sum_{j,k} \left(E(j) - \mu_{2,\mathrm{eq}}\right) T_2(j, t|k) P_{1,\mathrm{eq}}(k). \tag{3.42}$$

The l.h.s. of the equation is the deviation of the time dependent average energy from its equilibrium value in situation '2'. By expanding

$P_{1,\mathrm{eq}}$ in terms of $P_{2,\mathrm{eq}}$ to linear order in $\beta_1 - \beta_2$, we find

$$P_{1,\mathrm{eq}}(k) = P_{2,\mathrm{eq}}(k)\left(1 - (\beta_1 - \beta_2)(E(k) - \mu_{2,\mathrm{eq}})\right). \tag{3.43}$$

Exercise 14: Show Eq. (3.43) by writing $P_{1,\mathrm{eq}}(k)$ as $P_{\beta_2+(\beta_1-\beta_2),\mathrm{eq}}(k)$ and expanding to first order with respect to the small parameter $\beta_1 - \beta_2$.

Inserting Eq. (3.43) into Eq. (3.42), we also find

$$\begin{aligned} \Delta_E(t) &= -(\beta_1 - \beta_2)\sum_{j,k}\left(E(j) - \mu_{2,\mathrm{eq}}\right)\left(E(k) - \mu_{2,\mathrm{eq}}\right)T_2(j,t|k)P_{2,\mathrm{eq}}(k) \\ &= -(\beta_1 - \beta_2)\mathrm{Corr}_E(t) = -\frac{\mu_{2,\mathrm{eq}} - \mu_{1,\mathrm{eq}}}{\sigma^2_{2,\mathrm{eq}}}\mathrm{Corr}_E(t). \end{aligned} \tag{3.44}$$

The last equality follows from a linear expansion of the average equilibrium energy difference between the two situations, and from the relation

$$\frac{d\mu_{\mathrm{E}}(\beta)}{d\beta} = -\sigma^2_{\mathrm{E}}(\beta). \tag{3.45}$$

Exercise 15: The zero'th order term in the Taylor's expansion (3.43) does not contribute to Eq. (3.44). Show that

$$\sum_{j,k}\left(E(j) - \mu_{2,\mathrm{eq}}\right)T_2(j,t|k)P_{2,\mathrm{eq}}(k) = 0. \tag{3.46}$$

Exercise 16: The equilibrium average of the energy is

$$\mu_{\mathrm{E}}(\beta) = \frac{\sum_j E(j)e^{-\beta E(j)}}{\sum_j e^{-\beta E(j)}}. \tag{3.47}$$

Verify Eq. (3.45) by differentiating Eq. (3.47) with respect to β.

By adding $\mu_{2,\mathrm{eq}} - \mu_{1,\mathrm{eq}}$ to both sides of Eq. (3.44), one obtains the linear response function

$$R_E(t) = (\mu_{2,\mathrm{eq}} - \mu_{1,\mathrm{eq}})(1 - C_E(t)). \tag{3.48}$$

On the r.h.s. of the equation, C_E is the energy autocorrelation function normalized to one at time zero. Note that, according to Eq. (3.35), the response function approaches $\mu_{2,eq} - \mu_{1,eq}$ for large t. Using the equilibrium heat capacity

$$C_v = \frac{d\mu_E(T)}{dT}, \tag{3.49}$$

the response function can alternatively be written in terms of the initial and final temperatures, T_1 and T_2 as:

$$R_E(t) = (T_2 - T_1)C_v(1 - C_E(t)). \tag{3.50}$$

The correlation function appearing in Eq. (3.50) should rightly carry the subscript 2. However, since the factor $(T_2 - T_1)$ already appears in this equation, the error committed by replacing it with the energy correlation function in the unperturbed situation '1' is of second order in $(T_2 - T_1)$ and can be safely ignored. The response to the perturbation is thus linearly related to the energy correlation function in the unperturbed state.

Exercise 17: Detail why the distinction between the correlation functions in states 1 and 2 can be safely ignored in Eq. (3.50).

Only little effort is needed to generalize Eq. (3.48) to other situations, where the energy function previously defined in Eq. (3.39) includes for each state j an additive term of the form $-M(j)H$, i.e. where

$$\epsilon(j) = \beta E(j) = \beta(E_0(j) - M(j)H). \tag{3.51}$$

The real valued function M could represent the magnetic dipole moment in a given spatial direction, and the real number H is the component of an external magnetic field in the same direction. Other interpretations are also possible. In general, variables M and H whose product enters the energy function are called *conjugate*.

Exercise 18: Check that the equilibrium average of the magnetization

$$\mu_{M.eq}(H) = \frac{\sum_j M(j)e^{-\beta E(j)}}{\sum_j e^{-\beta E(j)}} \tag{3.52}$$

can also be written as

$$\mu_{M,eq}(H) = \frac{1}{\beta}\frac{\partial}{\partial H}\log(Z). \tag{3.53}$$

The equilibrium magnetic susceptibility is defined as

$$\chi_{M,eq} = \frac{\partial \mu_{M,eq}(H)}{\partial H}. \tag{3.54}$$

Check that the susceptibility for $H = 0$ obeys

$$\chi_{M,eq}(0) = \beta\sigma^2_{M,eq}, \tag{3.55}$$

where $\sigma^2_{M,eq}$ is the equilibrium variance of M.

The counterpart of Eq. (3.44) for a magnetic field change $H_1 \to H_2$, with constant β, can be obtained by suitably massaging the notation. The symbols $\mu_{1,eq}$ and $\mu_{2,eq}$ will be reused for the equilibrium value of the average magnetization in situations '1' and '2', respectively. Similarly, $\sigma^2_{2,eq}$ will be the variance of the magnetization in situation '2'. With these conventions, we immediately obtain

$$\begin{aligned}\Delta_M(t) &= \beta(H_1 - H_2)\sum_{j,k}(M(j) - \mu_{2,eq}) \\ &\quad \times (M(k) - \mu_{2,eq})T_2(j,t|k)P_{2,eq}(k) \\ &= \beta(H_1 - H_2)\mathrm{Corr}_M(t) = -\frac{\mu_{2,eq} - \mu_{1,eq}}{\sigma^2_{2,eq}}\mathrm{Corr}_M(t).\end{aligned} \tag{3.56}$$

Adding $\mu_{2,eq} - \mu_{1,eq}$ to both sides of Eq. (3.44) yields

$$R_M(t) = (\mu_{2,eq} - \mu_{1,eq})(1 - C_M(t)). \tag{3.57}$$

The magnetic response to a stepwise field change of size ΔH occurring at time zero can be rewritten using the equilibrium magnetic susceptibility $\chi_{M,eq}$ and the temperature T as

$$R_M(t) = \Delta H \chi_{M,eq}(1 - C_M(t)). \tag{3.58}$$

Again, C_M describes the equilibrium autocorrelation in situation '2', but can be replaced, to linear order in ΔH, with its counterpart in situation '1'. The two sides of Eq. (3.57) describe the dynamical properties of the perturbed and unperturbed system, respectively.

To summarize, if H is an externally adjustable parameter conjugated to a state space function M, Eq. (3.58) is the response to its stepwise increase. Formally, the time dependence of H is in this case $H(t'') = \Delta H \eta(t'')$, where η is the Heavyside step function. For a step occurring at a generic time $t' \neq 0$, the time dependence of H is $H(t'') = \eta(t'' - t')\Delta H$. By shifting the time variable, the response to this latter change is seen to be

$$R_M(t - t') = \Delta H \chi_{M,eq} \eta(t - t')(1 - C_M(t - t')). \tag{3.59}$$

Let $r_M(t - t')$ be the response at time t to an instantaneous kick $H(y) = \Delta H \delta(y - t')$, which is imposed at time t'. This response is then obtained by differentiating Eq. (3.59) with respect to t':

$$r_M(t - t') = -\frac{dR_M(t - t')}{dt'}. \tag{3.60}$$

Linear susceptibilities are always partial derivatives of response functions with respect to the parameter being changed. The susceptibility corresponding to an instantaneous kick, called *memory kernel*, is useful to determine the linear response to an arbitrary time dependent perturbation, $H(t)$. Using Eqs. (3.60) and (3.59), the memory kernel is given by

$$\kappa_M(t - t') = \chi_{M,eq} \eta(t - t') \frac{C_M(t - t')}{dt'}. \tag{3.61}$$

The linear response to $H(t')$ is expressed by the convolution:

$$R_M(t) = \int_{-\infty}^{t} H(t') \kappa_M(t - t') dt'. \tag{3.62}$$

The fact that the upper limit of the integral is t embodies *causality*: The future shape of $H(t)$ cannot affect the current form of the response. As an application of Eq. (3.62), a complex AC susceptibility

$\chi_{M,AC}(\omega)$ is defined as the response to a complex harmonically oscillating perturbation $Ae^{i\omega t'}$ divided by the perturbation itself:

$$\chi_{M,AC}(\omega) = \int_0^\infty \kappa_M(t')e^{-i\omega t'}dt'. \tag{3.63}$$

The AC susceptibility is usually written as

$$\chi_{M,AC}(\omega) = \chi'_M - i\chi''_M, \tag{3.64}$$

where χ'_M and χ''_M are called the in- and out of phase susceptibilities. Using Eqs. (2.18), (3.61), (3.63) and (3.55), we obtain

$$\chi_{M,AC}(\omega) = \chi_{M,eq}\left(1 - \omega\int_0^\infty C_M(t)\sin(\omega t)dt\right) - \frac{i\omega}{2k_BT}(\omega), \tag{3.65}$$

where S_M is the power-spectrum of the spontaneous magnetic fluctuations. We note that, in the limit $\omega \to 0$, the AC susceptibility approaches the equilibrium susceptibility. Secondly, the out of phase component is directly related to the power spectrum and therefore gauges energy dissipation.

3.4. RELAXATION PROPERTIES

In many important cases, a Markov process or chain relaxes to a stationary distribution. The stationary regime is important in Monte Carlo applications where, as we shall see, algorithms are specifically designed to reach stationarity as quickly as possible. The way in which equilibrium is reached or approached is highly relevant for physical applications. The existence of a stationary state coupled to requirements on the connectivity of the configuration space is enough to determine important general properties of the relaxation dynamics.

3.4.1. Relaxation to a Prescribed Stationary State

In physical applications, the stationary distribution P_{st} of a Markov process can usually be identified with an equilibrium distribution of the general form given in Eq. (3.38). When the number of states in configuration space N is finite, and the dynamics is ergodic, one can assume $P_{\text{st}}(k) > 0$, for all states k. The principle of detailed balance (discussed below) then guarantees convergence of the time dependent process to the required stationary state. However, in the thermodynamic limit, $N \to \infty$, convergence is not unconditionally guaranteed. Considerable interest is attached to *broken symmetry* situations, where the dynamics relaxes to a state different from the state predicted by equilibrium considerations. Convergence to a prescribed stationary state is also crucial in *importance sampling*. This is a Monte Carlo method based on Markov chains and widely used for the estimation of integrals (or sums) over multidimensional domains. The method is discussed in the next chapter.

The task ahead of us is thus to construct a Markov process, or Markov chain, converging toward a given stationary distribution. As we shall see, this task has multiple solutions. In importance sampling, this multiplicity provides the opportunity to optimize the dynamics for faster convergence. However, for Markov processes intended to model time dependent phenomena, the non-uniqueness of the dynamics clearly poses a problem. In this case, additional empirical evidence is needed to support any specific choice.

The rate coefficients

We restrict our attention to cases satisfying the *detailed balance* condition,

$$k_{ij} P_{\text{eq}}(j) = k_{ji} P_{\text{eq}}(i), \tag{3.66}$$

where $P_{\text{eq}} > 0$ is given in Eq. (3.38). One implication of detailed balance is that $k_{ij} \neq 0 \Rightarrow k_{ji} \neq 0$. Irreversible transitions between micro-states are thus forbidden. For physical systems, detailed balance is a consequence of the time reversal symmetry of the underlying microscopic dynamical equation. Noting that

$$\frac{k_{ij}}{k_{ji}} = \frac{P_{\text{eq}}(i)}{P_{\text{eq}}(j)} = e^{\epsilon(j) - \epsilon(i)}, \tag{3.67}$$

detailed balance constrains the form of the rate coefficients. It does, however, still leave a considerable freedom of choice, e.g. an arbitrary multiplicative factor can be associated to each pair of states (i,j). Clearly, then, Eq. (3.67) does not fix the absolute values of the time scales of the relaxation process described by the master equation.

Exercise 19: Assume that a set of positive quantities k_{ij} is given, allegedly representing the rates of a Markov process with detailed balance. Show that, if any two points in configuration space are connected by a unique path, a scalar function ϵ can be constructed fulfilling Eq. (3.67). Which type of configuration space has the required topology? Which additional conditions must be imposed on the k_{ij} if they are to represent transition rates in an arbitrary configuration space, where points are connected by multiple paths?

Exercise 20: Assuming Eq. (3.67), show that the distribution P_{eq} is indeed a stationary solution of the master equation (3.16).

Time dependent solution

Under rather general conditions, the time dependent solution of the master equation converges toward P_{eq}. To show this result, we *symmetrize* the master equation, i.e. we transform it into the equation

$$\frac{d\mathbf{Q}(t)}{dt} = \tilde{\mathbf{W}}\mathbf{Q}(t), \tag{3.68}$$

where $\tilde{\mathbf{W}}$ is a symmetric and negative definite matrix, and where the matrix $\mathbf{Q}$ has entries

$$Q_{jk}(t) = T_{jk}(t)P_{\text{eq}}^{-\frac{1}{2}}(j) = T_{jk}(t)Z^{\frac{1}{2}}e^{\epsilon(j)/2}. \tag{3.69}$$

The initial condition is

$$Q_{jk}(t=0) = \delta_{jk}P_{\text{eq}}^{-\frac{1}{2}}(k). \tag{3.70}$$

For notational simplicity, we write $Q_{jk}(t)$ as $Q_j(t)$ omitting the index of the initial state. Equation (3.16) implies

$$\frac{dQ_j(t)}{dt} = \sum_{i \neq j} k_{ji} e^{(\epsilon(j)-\epsilon(i))/2} Q_i(t) - Q_j(t) \sum_{i \neq j} k_{ij}. \tag{3.71}$$

The elements of the matrix $\tilde{\mathbf{W}}$ are thus given by

$$\tilde{W}_{ji} = k_{ji} e^{(\epsilon(j)-\epsilon(i))/2}, \quad \text{for } i \neq j \quad \text{and} \quad \tilde{W}_{jj} = -\sum_{i \neq j} k_{ij}. \tag{3.72}$$

Exercise 21: Using Eq. (3.67), show that the matrix $\tilde{W}$ is symmetric.

Exercise 22: Let $\mathbf{S} = \text{Diag}(P_{\text{eq}}^{1/2})$ be a square diagonal matrix, whose j'th diagonal entry equals $P_{\text{eq}}^{1/2}(j)$. Verify that Eq. (3.72) can also be written as

$$\tilde{\mathbf{W}} = \mathbf{S}^{-1}\mathbf{W}\mathbf{S}. \tag{3.73}$$

The sought convergence properties of the relaxation process depend on $\tilde{\mathbf{W}}$ being negative definite. By definition, this is the case if and only if, for an arbitrary vector $\mathbf{x}$,

$$(\mathbf{x}, \tilde{\mathbf{W}}\mathbf{x}) = \sum_{i,j} x_i \tilde{W}_{ij} x_j \leq 0. \tag{3.74}$$

Using the definition of the diagonal matrix elements $W_{ii} = \sum_{j \neq i} k_{ji}$, the r.h.s. of the above equation can be rewritten as

$$(\mathbf{x}, \tilde{\mathbf{W}}\mathbf{x}) = \sum_{i,j,(i \neq j)} k_{ji} e^{(\epsilon(j)-\epsilon(i))/2} x_i x_j - \frac{1}{2} \sum_{i,j,(i \neq j)} (x_i^2 k_{ji} + x_j^2 k_{ij}). \tag{3.75}$$

Applying the detailed balance condition to the last term, one obtains the desired inequality

$$(\mathbf{x}, \tilde{\mathbf{W}}\mathbf{x}) = -\frac{1}{2} \sum_{i \neq j} \left(x_i - x_j e^{(\epsilon(j)-\epsilon(i))/2} \right)^2, \quad k_{ji} \leq 0. \tag{3.76}$$

Exercise 23: Verify the steps leading to Eq. (3.76).

A standard theorem of linear algebra states that symmetric matrices, such as our $\tilde{\mathbf{W}}$, have a complete set of orthonormal eigenvectors, all with real eigenvalues. Moreover, since $\tilde{\mathbf{W}}$ is negative definite, the eigenvalues are negative or zero. A zero eigenvalue is present whenever the problem has a stationary solution. Let us arrange the eigenvalues in decreasing order, and denote them by greek letters, e.g. λ. The corresponding eigenvectors are denoted by ψ_λ. The j' entry of the eigenvector is $\psi_\lambda(j)$. Orthonormality of the eigenvectors,

$$(\psi_\lambda, \psi_\mu) = \delta_{\lambda\mu} \tag{3.77}$$

entails the *completeness relation*

$$\sum_\mu \psi_\mu(j)\psi_\mu(k) = \delta_{kj}. \tag{3.78}$$

The solution of Eq. (3.69) can now be written as

$$Q_{ik}(t) = \sum_\lambda \psi_\lambda(i)\psi_\lambda(k)e^{\lambda t}\left(P_{\text{eq}}(k)\right)^{-\frac{1}{2}}. \tag{3.79}$$

In terms of $\mathbf{T}$ this amounts to

$$T_{ik}(t) = e^{-\frac{1}{2}(\epsilon(i)-\epsilon(k))}\sum_\lambda \psi_\lambda(i)\psi_\lambda(k)e^{\lambda t}. \tag{3.80}$$

In the limit $t \to \infty$, all exponential terms corresponding to negative eigenvalues decay to zero, and the only non-vanishing term in the sum on the r.h.s. of the equation is the first one, which corresponds to $\lambda = 0$. On the l.h.s. of the equation, $T_{ik}(t) \to P_{\text{eq}}(i)$. Accordingly,

$$\psi_0 = \mathbf{P}_{eq}^{1/2}. \tag{3.81}$$

Exercise 24: Verify that $\mathbf{P}_{eq}^{1/2}$ is an eigenvector of $\tilde{\mathbf{W}}$ by direct insertion in the r.h.s. of Eq. (3.71).

The elements of the equilibrium eigenvector ψ_0 are all real and positive. This guarantees the uniqueness of the stationary solution: Assume the existence of two linearly independent equilibrium eigenvectors ψ_0 and ϕ_0. Without loss of generality, these can be assumed to be mutually orthogonal.[2] Since they are both non-negative, they necessarily have disjoint support. Since however ψ_0 is nowhere zero, we have reached a contradiction.

Exercise 25: Using

$$\forall k : \sum_i T_{ik}(t) = 1, \tag{3.82}$$

and the linear independence of the eigenvectors ψ_λ, show that, for $\lambda \neq 0$,

$$\sum_i e^{-\frac{1}{2}\epsilon(i)} \psi_\lambda(i) = 0. \tag{3.83}$$

Exercise 26: Using Eq. (3.73) show that $\tilde{\mathbf{W}}$ and $\mathbf{W}$ have the same eigenvalues. The eigenvectors of $\mathbf{W}$ are given by

$$\mathbf{W}\mathbf{S}\psi_\lambda = \lambda \mathbf{S}\psi_\lambda. \tag{3.84}$$

The eigenvalue expansion in Equation (3.80) contains as many terms as there are states in the problem's configuration space. Let λ_r be the largest non-zero eigenvalue of the problem. For times larger than the *relaxation time* $t_{\text{rel}} = -\frac{1}{\lambda_r}$, Eq. (3.80) is dominated by the largest two terms, and the relaxation is to a good approximation exponential. However, in the limit $N \to \infty$ the number of eigenvalues, which is equal to N, goes to infinity. A possible scenario is that the eigenvalues become dense near zero, i.e. that the difference between contiguous eigenvalues in that region goes to zero. In this case, the final exponential stage of the relaxation process may become unobservable, since many nearly equal eigenvalues are equally important.

[2] If not, two linear combinations which are orthogonal can be obtained using Gram–Schmidt orthogonalization.

3.4.2. Eigenvalue Expansions of the Moments

In a stationary process, eigenvalue expansions for the moments are easily derived from Eq. (3.80). Consider a function F with zero average. Adapting the notation of Eq. (3.34), its correlation function is given by

$$\mathrm{Corr}_F(t) = \sum_{ik} F(i)F(k)T_{ik}(t)P_{\mathrm{eq}}(k). \tag{3.85}$$

Rewriting Eq. (3.80) as

$$T_{ik}(t) = \frac{1}{P_{\mathrm{eq}}(k)} \sum_{\lambda} P_{\mathrm{eq}}^{1/2}(i)\psi_\lambda(i)P_{\mathrm{eq}}^{1/2}(k)\psi_\lambda(k)e^{\lambda t} \tag{3.86}$$

we arrive at

$$\mathrm{Corr}_F(t) = \sum_{\lambda<0}\left(\sum_i F(i)P_{\mathrm{eq}}^{1/2}(i)\psi_\lambda(i)\right)^2 e^{\lambda t} \stackrel{\mathrm{def}}{=} \sum_{\lambda<0} \rho(\lambda)e^{\lambda t}. \tag{3.87}$$

Note that all $\rho(\lambda)$ are positive, and that the correlation hence decays in a monotonic fashion.

> **Exercise 27:** Explain why the term corresponding to the zero eigenvalue is absent from Eq. (3.87). *Hint*: the fact that F has zero average in equilibrium is of importance.

For large system size N, the eigenvalue spectrum often approaches a continuum. Using the notation $\rho(\lambda)d\lambda > 0$ for the part of the sum in Eq. (3.87) which spans the interval $\lambda \pm d\lambda/2$ in the eigenvalue spectrum, the eigenvalue expansion for the correlation function takes the form

$$\mathrm{Corr}_F(t) = \int_{\lambda<0} \rho(\lambda)e^{\lambda t}d\lambda. \tag{3.88}$$

The correlation function now appears as the Laplace transform of the spectral density $\rho(\lambda)$. In the lack of further information, little can be said theoretically on the time dependence of the correlation function, apart from excluding the possibility of oscillatory behaviour. Fortunately, power spectra of experimental signals help to characterize the respective correlation functions, see, e.g., Eqs. (2.18) and (2.20).

Example 3: According to Eq. (2.16), the power spectrum of a stationary stochastic process is the Fourier transform of the corresponding autocorrelation function. Very often, experimental power spectra have the general form

$$S(\omega) = \frac{2A\tau}{1 + (\tau\omega)^2}, \tag{3.89}$$

where the constant A provides the correct dimension. Since the first term in the denominator can usually be neglected, this form is called $1/f^2$ noise where $f = \omega/(2\pi)$ is the cyclic frequency. The form implies an exponential decay, $C(t) = e^{-t/\tau}$, of the correlation function. Naively, one could infer from our previous discussion that the eigenvalue spectrum of the Markov process describing the fluctuations has only one non-zero eigenvalue, i.e. $\lambda = -1/\tau$. In reality, numerous eigenvalues may exist but be unobservable, either because the corresponding modes decay too quickly or because they remain frozen throughout the observation time window.

3.5. MARKOV CHAINS

Markov chains are the workhorse of computer simulations of all kinds. As mentioned, they are simply Markov processes with discrete time. The following discussion of their relaxation properties can thus borrow from our treatment of the master equation.

The dynamical equations, Eq. (3.18), can be written compactly as the matrix equation

$$\mathbf{T}(t+1) = \mathbf{M}\mathbf{T}(t), \tag{3.90}$$

which is the discrete counterpart to Eq. (3.21). The solution of the Markov chain with initial condition $\mathbf{T}(0) = \mathbf{I}$ has the form

$$\mathbf{T}(t) = \mathbf{M}^t. \tag{3.91}$$

Each matrix element M_{ij} represents the probability of moving from state j to i in a single time step. Conservation of probability entails $\sum_i M_{ij} = 1$, for all j, which amounts to **M**'s column sums being all equal to one. In many cases, convergence of the Markov chain to a prescribed stationary distribution $P_{\text{eq}}(j) = \frac{e^{-\epsilon(j)}}{Z}$ is desired. This property is ensured by the detailed balance condition

$$M_{ij} = e^{\epsilon(j)-\epsilon(i)} M_{ji}, \tag{3.92}$$

a condition also implying

$$\mathbf{M} P_{\text{eq}} = P_{\text{eq}}. \tag{3.93}$$

The equality follows from Eq. (3.91) together with the equation

$$M^t_{ij} = e^{\epsilon(j)-\epsilon(i)} M^t_{ji}, \quad t = 1, 2, 3 \ldots, \tag{3.94}$$

where M^t_{ij} denotes entry ij of the matrix $\mathbf{M}^t$. A convenient restatement of detailed balance as a matrix equation is as follows: Let **D** be a diagonal matrix with $D_{jj} = P_{\text{eq}}(j)$, and let the superscript T denote transposition. The stochastic matrix **M** fulfils detailed balance if and only if

$$\mathbf{D}\mathbf{M}^{\mathrm{T}}\mathbf{D}^{-1} = \mathbf{M}. \tag{3.95}$$

The latter equation implies that for any integer t, $\mathbf{M}^t$ fulfills detailed balance.

Exercise 28: Check Eqs. (3.94), (3.92) and (3.95) .

Define now **W** by the matrix equation

$$\exp(\mathbf{W}) = \mathbf{M}. \tag{3.96}$$

Expanding the exponential function in its power series, and applying Eq. (3.95) to both sides of the equation, it is shown that if **M** satisfies detailed balance, **W** does as well. The column sums of **W** and of all its powers are seen to vanish. Hence, **W** is a stochastic matrix of the type earlier discussed, and the limit $\exp(\mathbf{W}t) = \mathbf{M}^t$ exists for $t \to \infty$. In this limit, each column of **M** approaches the stationary distribution P_{eq}, which, as earlier discussed, is unique in a finite configuration space. Comparison with Eq. (3.91) shows that the Markov chain converges as expected. It is clear from Eq. (3.96) that **W** and **M** have the same eigenvectors. Furthermore, in the basis where **W** is diagonal, the only non-zero elements are precisely the eigenvalues. Since these are non-positive, it follows that the eigenvalues of **M** lie in the interval $(0, 1]$.

3.6. SIX EXAMPLES OF MASTER EQUATIONS

To get acquainted with the basic properties of master equations, we consider six examples: The Poisson process, the walk on a fully connected set of states, the random walk in Euclidean space, the random walk on a finite set of states, the birth and death process and the dichotomic process.

The Poisson process

The states of the Poisson process are the non-negative integers $n \geq 0$. Non-zero rate coefficients are all identical, and connect each state n with its neighbour state $n + 1$. Omitting for convenience the index n_0 denoting the initial condition, the master equation reads

$$\frac{dT_n(t)}{dt} = \kappa(T_{n-1}(t) - T_n(t)). \tag{3.97}$$

The solution must decay fast enough at infinity for T to be summable, since conservation of probability requires

$$\sum_{n=n_0}^{\infty} T_n(t) = 1. \tag{3.98}$$

Exercise 29: Show that the solution of Eq. (3.97) is

$$\begin{aligned} T_{nn_0}(t) &= \frac{(\kappa t)^{n-n_0}}{(n-n_0)!} e^{-\kappa t}, \quad n \geq n_0, \\ T_{nn_0}(t) &= 0, \quad n < n_0. \end{aligned} \tag{3.99}$$

Exercise 30: Find the expectation value $\mu(t)$ and the variance $\sigma^2(t)$ of the Poisson process *i)* using the explicit solution given above and *ii)* directly from Eq. (3.97).

Exercise 31: A Poisson process is in state n_0 at time t_0. Let the 'survival' probability $P(t)$ be the probability that no change of state or 'jump' has occurred at time $t > t_0$. Write down and solve the equation for the survival probability. The probability of survival for a time $\delta t = t - t_0$ does not depend on t_0. For this reason, the Poisson process is called 'memoryless'. Can the probability of a person's survival be modelled as a Poisson process?

Exercise 32: Find at least two examples of natural processes which are Poissonian.

Clearly, the Poisson process is homogeneous but not stationary. Its solution 'runs off to infinity' as time grows.

Random walks on a fully connected graph

The state space of the Poisson process is both simple, and sparsely connected. An equally simple case at the opposite end of the connectivity scale is a fully connected graph with N states, labelled from 0 to $N-1$. Any two states are connected by an edge, and the rate coefficients are all equal to κ. The master equation reads

$$\frac{dT_n(t)}{dt} = \kappa\left(\sum_{k\neq n} T_k(t) - T_n(t)(N-1)\right), \quad n = 0, 1, \ldots N-1 \tag{3.100}$$

with the initial condition

$$T_n(t=0) = \delta_{n,0}. \tag{3.101}$$

Exercise 33: Using the normalization condition $\sum_{k\neq n} T_k(t) = 1 - T_n(t)$, rewrite Eq. (3.100) into a set of identical and un-coupled equations. Show then that

$$T_n(t) = \frac{1}{N}\left(1 - e^{-\kappa N t}\right) \quad \text{for } n \neq 0, \tag{3.102}$$

and

$$T_0(t) = \frac{1}{N}\left(1 - e^{-\kappa N t}\right) + e^{-\kappa N t}. \tag{3.103}$$

A uniform stationary distribution is approached on a single time scale, the relaxation time $t_{\rm rel} = 1/(\kappa N)$. The relaxation time decreases with N, an unusual situation reflecting the full connectivity of the configuration space. Fully connected configuration spaces are not usually encountered in applications. However, fully connected subsets of states in configuration space, so called *cliques*, are sometimes present in configuration space.

3.6.1. Random Walks in Euclidean Space

Stochastic processes of the type

$$S(n) = \sum_{k=1}^{n} d_k,$$

where the increments d_k are independent and identically distributed random variables are simple examples of Markov chains. The nature of their configuration space depends on that of the increments, which could be reals, integers or vectors. The distribution P_d of the increments determines the connectivity of the configuration space, e.g. states x and y are connected if and only if $P_d(y - x)$ differs from zero. If $P_d > 0$ for all x and y, configuration space is fully connected. However, unlike the previous examples, not all transitions along the connections are equiprobable. The propagator can be expressed as a simple convolution

$$T(x, n) = \int P_d(x - y)T(y, n - 1)dy, \tag{3.104}$$

where the integral extends over the support of P_d. The above equation is easily solved using Fourier or Laplace transforms, which both map convolutions into products. Some special cases can be dealt with more directly: E.g. when P_d is Gaussian, S is a sum of independent Gaussian variables, and hence itself a Gaussian. If, furthermore, the increment has zero average and unit variance, the only free parameter left to characterize S is its variance.

An important example in this class is the random walk in d-dimensional Euclidean space. At each tick of the clock, the walker picks one of the available directions with equal probability, say direction i, and changes her position by an amount Δr_i in that direction. All Δr_i are statistically independent and identically distributed Gaussian variables of zero mean and unit variance. With n_i steps taken in direction i, the displacement $r_i(n_i)$ is hence a Gaussian variable with zero expectation and variance $\sigma_i^2(n_i) = r_i^2(n_i) = n_i$. The squared average distance from the current position to the origin is $r^2 = \sum_{i=1}^d r_i^2 = \sum_{i=1}^d n_i$. Noting that the total number of steps is $n = \sum_{i=1}^d n_i$, we find

$$r^2(n) = n. \tag{3.105}$$

Introducing physical time and length units, each step has variance $2D\delta t$, where D is the diffusion coefficient, and where δt is the waiting time between consecutive steps. The squared distance from the origin at time $t = n\delta t$ is then

$$r^2(t) = 2Dt. \tag{3.106}$$

This result is a mathematical consequence of the additivity of the variance of the independent increments of the walk. Its ubiquity simply reflects a lack of structure. Physically, it highlights that diffusion is a highly inefficient transport mechanism over large distances.

Exercise 34: Suppose that the time dependent position $r(t)$ of a particle is, for very short time scales, described by a known stochastic process, which is not itself a random walk. Which statistical property of $r(t)$ identifies a time scale δt such that a random walk description could be appropriate for $t \gg \delta t$?

Random walks on a finite set

Boundaries remove translational invariance, and introduce a higher level of complexity. Consider a random walk on a finite lattice in one dimension, with points labelled $0, 1, \ldots N-1$. Except near the boundaries, the walker moves at a constant rate κ, either to the left or the right with equal probability.

Depending on the boundary conditions, different interpretations are possible. For *absorbing* boundary conditions, we reencounter the gambler's ruin problem: Each move to the right or to the left corresponds to winning or losing a bet. The position represents the gambler's capital, and the game stops at positions 0 and N. Either the gambler or his opponent has then lost all his capital. *Reflecting* boundary conditions describe a confined particle which turns back upon reaching a boundary.

In the interior of the domain, the master equation reads

$$\frac{dT_n(t)}{dt} = \kappa(T_{n-1}(t) + T_{n+1}(t) - 2T_n(t)), \quad 0 < n < N, \tag{3.107}$$

with the initial condition

$$T_n(t=0) = \delta_{n,n_0}. \tag{3.108}$$

In the absorbing case, the probability flow into state 1 from the left and state $N-1$ from the right must vanish. This is accommodated by the boundary condition

$$T_N(t) = 0; \quad T_0(t) = 0. \tag{3.109}$$

In the reflecting case, the *net* probability flow in states 0 and N must vanish:

$$\frac{dT_N}{dt} = 0; \quad \frac{dT_0}{dt} = 0. \tag{3.110}$$

The two versions of the problem can be treated together using the even and odd periodic solutions to Eq. (3.107). The functions are given by

$$E_k(n) = \left(\frac{1}{N}\right)^{\frac{1}{2}} \cos\left(\frac{k\pi n}{N}\right), \quad 0 < k < N \tag{3.111}$$

$$E_N(n) = \left(\frac{1}{2N}\right)^{\frac{1}{2}} (-1)^n, \tag{3.112}$$

$$E_0(n) = \left(\frac{1}{2N}\right)^{\frac{1}{2}}, \tag{3.113}$$

$$O_k(n) = \left(\frac{1}{N}\right)^{\frac{1}{2}} \sin\left(\frac{k\pi n}{N}\right), \quad 1 < k < N. \tag{3.114}$$

For each possible value of k, they fulfil the eigenvalue equation

$$\Delta^2 \mathbf{E}_k = \lambda_k \mathbf{E}_k \tag{3.115}$$

$$\Delta^2 \mathbf{O}_k = \lambda_k \mathbf{O}_k, \tag{3.116}$$

where Δ^2 is the one-dimensional lattice Laplacian, $(\Delta^2 f)(n) = f(n-1) + f(n+1) - 2f(n)$. The eigenvalues are

$$\lambda_k = 2\left(\cos(\frac{\pi k}{N}) - 1\right) = -4\sin^2\left(\frac{\pi k}{2N}\right). \tag{3.117}$$

They all lie in the interval $(-4, -0]$, and, except for $k = 0$ and $k = N$, they are doubly degenerate.

The solution of Eq. (3.110) on the set of points $1 \ldots N-1$, with absorbing boundary conditions, Eq. (3.109) and with sharp initial conditions at n_0 is given in terms of the sine series as

$$T_{n,n_0} = 2\sum_{k=1}^{N-1} O_k(n) O_k(n_0) e^{\kappa \lambda_k t}. \tag{3.118}$$

Correspondingly, the solution with reflecting boundary condition is given in terms of the cosine series by

$$T_{n,n_0} = 2\sum_{k=0}^{N-1} E_k(n) E_k(n_0) e^{\kappa \lambda_k t}. \tag{3.119}$$

Exercise 35: Check that Eqs. (3.118) and (3.119) satisfy the required equations and initial and boundary conditions.

In the absorbing case, the stationary solution is identically zero. The gambler will therefore eventually reach states 0 or N, either losing all his capital, or winning all of his opponent's capital. With reflecting boundary condition, the stationary solution is a proper equilibrium distribution, with all states equiprobable. In both cases the final state is approached on a time scale proportional to minus the inverse of the largest non-zero eigenvalue. This so-called relaxation time is

$$\tau_{\text{rel}} = -\frac{1}{\kappa\lambda_1} = \frac{1}{4\kappa \sin^2(\frac{\pi}{2N})}. \tag{3.120}$$

The relaxation time diverges as N^2 in the limit of large N.

> **Exercise 36:** If the lattice size N is very large, and the initial position of the walker is far from the boundaries, the relaxation process will initially resemble free diffusion on a line. By matching length scales, argue qualitatively that the effects of the boundaries becomes appreciable on a time scale
>
> $$\tau_b = \frac{N^2}{\kappa}. \tag{3.121}$$
>
> How does this time scale compare with the relaxation time?

Birth and death process

In a certain population, individuals reproduce or die at rates b_r and d_r, respectively. The rates of population change are of course proportional to the number of individuals present. The master equation then reads

$$\begin{aligned}
\frac{dT_n(t)}{dt} &= b_r(n-1)T_{n-1}(t) + d_r(n+1)T_{n+1}(t) \\
&\quad - (b_r + d_r)nT_n(t); \quad n \geq 1 \\
\frac{dT_0(t)}{dt} &= d_r T_1(t),
\end{aligned} \tag{3.122}$$

with the initial condition

$$T_n(t=0) = \delta_{n,n_0}. \tag{3.123}$$

By stipulating that $T_n(t) = 0$ for $n < 0$, Eq. (3.122) is automatically satisfied. The state with zero population has no loss term, and provides a trivial stationary state of the problem. This stationary state is attractive, or absorbing, for $b_r < d_r$.

Equation (3.122) can be solved exactly. However, it is more instructive at this stage to derive equations for the first two moments, μ and σ^2 of the distribution $T(n,t)$. We note in passing that closed form equations for the moments can only be derived if the rate coefficients of the master equation are linear functions of the independent variable, as is the case in the present application. Multiplying Eq. (3.122) by n and summing over all n, one obtains

$$\frac{d\mu(t)}{dt} = (b_r - d_r)\mu(t). \tag{3.124}$$

Analogous steps lead to an equation for the time dependence of the variance:

$$\frac{d\sigma^2(t)}{dt} = 2(b_r - d_r)\sigma^2(t) + (b_r + d_r)\mu(t). \tag{3.125}$$

The initial conditions for the two equations are $\mu(0) = n_0$ and $\sigma^2(0) = 0$.

> **Exercise 37:** Solve Eqs.(3.124) and (3.125). For $b_r < d_r$ both average and variance go to zero with increasing t, while for $b_r > d_r$ they both diverge. Clearly, the model in not adequate to describe a steady state situation. Keeping the same connectivity, how could the rate coefficients be changed for the equations to admit a non-trivial stationary state?

The dichotomic process

A dichotomic process has only two states, and analytical solutions are easily constructed. Dichotomic processes are widely used in the physical literature, where they are known as two level systems.

The two level system considered has states 1, the ground state, and 2 the excited state, with energies $E_1 = 0$ and $E_2 = \epsilon$. Furthermore a spin variable is defined as $\sigma_1 = 1$ and $\sigma_2 = -1$. The dynamics is a Markov chain whose transition rates $\kappa_{ij},\ i,j \in \{1,2\}$ obey detailed balance. We calculate the spin autocorrelation function

$$\mathrm{Corr}_\sigma(t) = \sum_{i=1}^{2}\sum_{j=1}^{2} P_{\mathrm{eq}}(i)T_{ji}(t)(\sigma_i - m)(\sigma_j - m), \tag{3.126}$$

where

$$m \stackrel{\mathrm{def}}{=} \frac{1 - e^{-\epsilon/T}}{1 + e^{-\epsilon/T}} = \tanh(\epsilon/2T) \tag{3.127}$$

is the thermal equilibrium average of the spin variable, at temperature T. The equilibrium distribution for the problem is

$$\mathbf{P}_{\mathrm{eq}} = \frac{1}{1 + e^{-\epsilon/T}} \begin{pmatrix} 1 \\ e^{-\epsilon/T} \end{pmatrix} \tag{3.128}$$

and the propagator (transition probability matrix) is

$$\mathbf{T}(t) = \frac{1}{1 + e^{\epsilon/T}} \begin{pmatrix} e^{\epsilon/T} + e^{-t/\tau} & e^{\epsilon/T}(1 - e^{-t/\tau}) \\ 1 - e^{-t/\tau} & 1 + e^{\epsilon/T}e^{-t/\tau} \end{pmatrix}, \tag{3.129}$$

where the relaxation time τ is given by

$$\tau = \frac{1}{\kappa_{12} + \kappa_{21}}. \tag{3.130}$$

Exercise 38: Write down the master equation for the dichotomic process. Check then that the transition probabilities given in Eq. (3.129) solve this master equation, have the correct initial values and have the correct asymptotic behaviour for $t \to \infty$.

To calculate the correlation function we note that

$$\sigma_{\pm 1} - m = \frac{e^{\pm\epsilon/2T}}{\cosh(\epsilon/2T)}.$$

Using Eq. (3.126), we arrive at

$$\mathrm{Corr}_\sigma(t) = \frac{e^{-t/\tau}}{\cosh^2(\epsilon/2T)}. \tag{3.131}$$

Exercise 39: Show that $\mathrm{Corr}_\sigma(t=0) = \mathrm{var}(\sigma)$.

Exercise 40: Find the (two) eigenvalues of the **W** matrix of the master equation for the dichotomic process, and derive the correlation function using this result.

3.7. CONTINUOUS TIME RANDOM WALKS

Continuous Time Random Walks were introduced in the physics literature by Scher and Montroll [SM75] as models for anomalous transport in amorphous semiconductors. The CTRM is a Markov chain $x(n)$, where the (discrete) time variable n is itself a stochastic process $n(t)$, e.g. we can imagine x being controlled by an unreliable clock, where the waiting time between successive 'ticks' is random. As a consequence, the number of ticks $n(t)$ in the time interval $(0, t)$ is a stochastic process. The process x is then said to be *subordinated* to n.

3.7.1. Subordination

We refer the reader to Feller [Fel66] for a more thorough mathematical discussion of interesting issues related to subordination, and start our discussion with an illustrative example.

Example 4: A fire station serves a certain area of Gotham City. Assume that the station's fire truck logs d_k miles, the distance to the location of the fire, at the k'th fire of the fiscal year. After n fires, the fire truck has hence logged

$$x(n) = \sum_{k=1}^{n} d_k$$

miles. Let us assume that the various distances d_k are independent and identically distributed variables, sharing the common density

$$P_d(z) = ae^{-az}; \quad 0 < z < \infty.$$

As the sum of n exponentially distributed variables, $x(n)$ is Gamma distributed, the probability density of logging x miles after

n fires is

$$T(x,n) = a\frac{(ax)^{n-1}}{(n-1)!}e^{-ax} \quad n = 1,2,\ldots. \tag{3.132}$$

This can be interpreted as the conditional probability density for mileage x, given that n fires have occurred.

For maintenance and budget purposes, one is more interested in mileage versus time. Let $P(n,t)$ be the probability that n fires occur in the time interval $(0,t)$, and let $M(x,t)$ be the probability density of logging x miles in time t. The latter is then given by

$$M(x,t) = \sum_{n=1}^{\infty} T(x,n)P(n,t) + P(0,t)\delta(x), \tag{3.133}$$

where the term containing the δ function accounts for the possibility that no fires occur.

Often, $n(t)$ can be treated as a Poisson process. In this case, we find

$$M(x,t) = e^{-\lambda t}\left(\sum_{n=1}^{\infty} T(x,n)\frac{(\lambda t)^n}{n!} + \delta(x)\right). \tag{3.134}$$

Evaluating the sum gives

$$M(x,t) = e^{-(ax+\lambda t)}\left(\frac{\lambda t a}{x}\right)^{\frac{1}{2}} I_1\left((4ax\lambda t)^{1/2}\right) + e^{-\lambda t}\delta(x), \tag{3.135}$$

where I_1 is a modified Bessel function of the first kind.

In the example above, the cost of operating the fire truck depends on the number of fires n, which is therefore the most natural variable to use in the description. In more realistic cases, time series and other observational data are to be modelled. If a subset of 'salient' events, corresponding to the fires, can be identified from the data as the events driving the dynamics, the resulting description can be hugely simplified. Finding a small set of events to which a dynamical process is subordinated reduces the complexity of the description and can be of great help in a modelling effort. The Continuous Time Random Walks (CTRW) discussed below are an example of this type of modelling approach, where salient events are identified as jumps from one meta-stable situation to another.

3.7.2. Scaling Properties of CTRW

As in the fire truck example, an integer variable $n(t)$ represents the number of steps in a Markov chain which occur in the time interval $(0, t)$. We imagine that the different states of this chain represent *traps* in the configuration space of the problem at hand. All these traps are dynamically equivalent and are described by the same probability density $W(t)$ that a trajectory trapped at time zero leaves the trap at time t. After each jump, the memory of previous states is erased, and the system starts afresh in its new state. Discussed below are the time dependences of the average and the variance of the number of jumps in CTRW models and how they scale with system size. Secondly, we discuss how the dynamical properties of the Markov chain in the n variable appear in the time domain.

Leaving aside, for the moment, the nature of the Markov chain, which of course depends on the problem at hand, we focus on the probability $P(n, t)$ that precisely n jumps occur in time t. If n events have occurred at time t, $n-1$ events must have occurred at an earlier time t'. Hence,

$$P(n,t) = \int_0^t P(n-1,t')W(t-t')dt'; \quad n = 1, 2 \ldots \tag{3.136}$$

$$P(0,t) = 1 - \int_0^t W(t-t')dt'. \tag{3.137}$$

The first equation is a convolution, which is mapped into a product by the Laplace transform $f \to \mathcal{L}_f$, where

$$\mathcal{L}_f(s) = \int_0^\infty e^{-st} f(t)dt. \tag{3.138}$$

With mild regularity conditions on f, the integral is defined in the complex s plane for $\mathrm{Re}(s) \geq 0$.

Exercise 41: Let $f \bullet g$ denote the convolution of the two functions f and g:

$$f \bullet g(t) = \int_0^t f(t')g(t-t')dt'. \tag{3.139}$$

Show that

$$\mathcal{L}_{f \bullet g}(s) = \mathcal{L}_f(s)\mathcal{L}_g(s). \tag{3.140}$$

Exercise 42: Assuming that $\mathcal{L}_W(s)$ is differentiable at $s = 0$, show that the average waiting time τ_0 is given by

$$\tau_0 = -\frac{d}{ds}\mathcal{L}_W(s)_{|s=0}. \tag{3.141}$$

Applying the Laplace transform to Eq. (3.137), we find

$$\mathcal{L}_P(n,s) = (\mathcal{L}_W(s))^n \frac{1 - \mathcal{L}_W(s)}{s}. \tag{3.142}$$

The above formula can be applied to obtain the Laplace transform of the first two moments of $P(n,t)$. The first moment, $\overline{n(t)} = \sum_0^\infty nP(n,t)$ has Laplace transform

$$\mathcal{L}_{\overline{n}}(s) = \sum_{n=0}^{\infty} n\mathcal{L}_P(n,s) = \frac{\mathcal{L}_W(s)}{s(1 - \mathcal{L}_W(s))}. \tag{3.143}$$

Exercise 43: Using that

$$\sum_{n=0}^{\infty} n\,(\mathcal{L}_W)^n = \mathcal{L}_W \frac{d}{d\mathcal{L}_W}\frac{1}{1 - \mathcal{L}_W} \tag{3.144}$$

verify Eq. (3.143).

For the second moment, it is convenient to look at the auxiliary quantity $\overline{n^2 - n}$, which has Laplace transform

$$\mathcal{L}_{\overline{n^2-n}}(s) = \sum_{n=0}^{\infty} n(n-1)\mathcal{L}_P(n,s) = \frac{2}{s}\left(\frac{\mathcal{L}_W(s)}{1 - \mathcal{L}_W(s)}\right)^2. \tag{3.145}$$

We now derive the long-time asymptotic behaviour of the average and variance of n. First, we consider an exponentially decaying $W(t)$, a form which is a suitable approximation when the average waiting time in a trap is finite. In this case it is easy to see that $n(t)$ is a Poisson process.

Exercise 44: Find the Laplace transform of $W(t) = \exp(-t/\tau_0)$. Use your result and Eq. (3.142) to show that $P(n,t)$ is a Poisson process.

For a general form of $W(t)$ it is not possible to construct closed form analytic formulas for $P(n,t)$. It is however possible to find the asymptotic long-time behaviour of $P(n,t)$ given a small s expansion of $\mathcal{L}_W(s)$. As shown by Eq. (3.141), a finite average waiting time is tantamount to $\mathcal{L}_W(s)$ being differentiable at $s = 0$. Hence, the small s expansion which we consider in the following,

$$\mathcal{L}_W(s) = 1 - (\tau_0 s)^x + \mathcal{O}(s^{x+1}), \quad 0 < x \leq 1 \tag{3.146}$$

implies an infinite average waiting time for $x < 1$.

The first term on the r.h.s. simply reflects the normalization of the waiting time distribution, i.e. that a jump will eventually occur with probability one. If $x = 1$, the derivative of the second term remains finite in the limit $s = 0$. In the time domain, and for $t \gg \tau_0$, $W(t)$ decays exponentially in time. Since the expectation value of $W(t)$ is given by $-\frac{d\mathcal{L}_W(s)}{ds}\big|_{s=0}$, τ_0 is in this case the average time spent in a trap. As mentioned, if $x < 1$ the average time spent in a trap is infinite. In this case, τ_0 simply represents a time scale needed for dimensional reasons. In the time domain, $W(t) \propto t^{-x-1}$ for $t \gg \tau_0$.

For any value of x, keeping only the leading terms in Eq. (3.143) we find that the average number of events at time t is

$$\overline{n}(t) = \frac{1}{x\Gamma(x)} \left(\frac{t}{\tau_0}\right)^x, \tag{3.147}$$

where Γ is the Gamma function. For $x = 1$, the expression reduces to $\overline{n}(t) = t/\tau_0$, as one would expect. For $x < 1$ the growth of $\overline{n}(t)$ is sub-linear in time. Applying the same procedure to Eq. (3.145), and using $\sigma_n^2(t) = \overline{n^2} - \overline{n}^2$, we finally obtain for the variance

$$\sigma_n^2(t) = \overline{n}(t) + \left(\frac{t}{\tau_0}\right)^{2x} \left(\frac{1}{x\Gamma(2x)} - \frac{1}{\Gamma^2(x+1)}\right). \tag{3.148}$$

The rightmost term on the r.h.s. vanishes and $\sigma_n^2(t) = \overline{n}(t)$, if and only if $x = 1$.

Since the equality of variance and expectation value is an important characteristic of the Poisson distribution, we expect that, for $t \gg \tau_0$, the statistics of the number of events will be reasonably well modelled by a Poisson distribution whenever the average waiting time is finite. Importantly the sum of two Poisson distributed variables remains Poisson distributed. Returning to the fires of Gotham city, enlarging the district with a new city means that the total number of fires is the sum of the number occurring in each district. Assuming that fires occur in an independent fashion in geographically distinct areas, the statistics of the total number of fires remains Poissonian, with a reduced τ_0. More generally, the number of fires will on average scale linearly with the number of districts included, and so will the variance. The linear scaling of the variance only reflects the statistical independence of fires in different subareas of the system.

The choice $x \neq 1$ is usually made in CTRW, since it leads to sub-linear growth in time of the number of events, and, correspondingly, to an 'anomalous' scaling of any physical quantity associated to these events. With this choice, the second term on the r.h.s. of (3.148) is dominant for $t \gg \tau_0$, and the variance becomes proportional to the square of the average. Hence, the description does not allow the average and variance of the number of events to be proportional to each other. This is however a natural requirement in systems composed of many dynamically independent parts, each supporting its own series of events.

We recall that the n 'jumps' of the CTRW constitute a Markov chain, which can pertain to a wide range of physical situations, e.g. random walks on a d-dimensional Euclidean lattice, on a fractal or on a tree graph. Let us then consider how the dynamical properties of this Markov chain come through in the time domain. In all cases, the n dependence of the solution is given by a linear combination of a number of exponentially decaying modes $e^{\lambda n}$, where $\lambda < 1$. The behaviour in the time domain arises by averaging with respect to

$P(n, t)$, an average which can be carried out independently for each mode. Mainly, we need therefore to assess the time dependence of the average of an arbitrary mode with respect to n. Let the latter quantity be denoted by $m(\lambda, t)$, and let $\mathcal{L}_m(\lambda, s)$ be the corresponding Laplace transform. We find

$$\mathcal{L}_m(\lambda, s) = \sum_{n=0}^{\infty} \mathcal{L}_P(n, s) e^{\lambda n} = \frac{1 - \mathcal{L}_W(s)}{s(1 - e^{\lambda}\mathcal{L}_W(s))}, \tag{3.149}$$

where we have used Eq. (3.142). Assume first that $W(t)$ has a finite average τ_0. Using Eq. (3.146) with $x = 1$, equation (3.149) can be expanded to lowest order in s to show that the mode decays exponentially in time, with a time scale $\tau_0/(1 - e^{\lambda})$. The time scale of the decay diverges for $\lambda \to 0$, precisely as one would expect. We also note that each λ contributes in a different way to the decay. Hence, the eigenvalue spectrum of the Markov chain enters the properties of the solution in the time domain.

The above results all follow because the s term in the denominator is cancelled by the term $\tau_0 s$ stemming from the expansion of the nominator. The situation is different when W has a long time tail, and when, correspondingly, $\mathcal{L}_W(s) = 1 - (\tau_0 s)^x$ with $x < 1$. To leading order, Eq. (3.149) gives a term proportional to s^{x-1}. This translates into a power-law decay in the time domain, with a tail exponent $-x$. In contrast to the previous result, the behaviour is independent of the value of λ. The dynamical properties of the jumps which are encoded in the eigenvalues of the Markov chain are hence, asymptotically for long times, washed away by the CTRW approach.

3.8. FOKKER–PLANCK EQUATIONS

Fokker–Planck (FP) equations are partial differential equations describing a class of time homogeneous Markov processes with a continuous state space, e.g. the Euclidean vector space in three dimensions. They control the time development of the propagator $T_2(x_2, t_2|x_1, t_1)$. Omitting as usual, the second spatial argument x_1, the subscripted index of T_2, and using boldface symbols for vectors and an additional overbar for tensors, the FP equations are written as

$$\frac{\partial T(\mathbf{r},t)}{\partial t} = -\nabla \cdot \mathbf{J}, \tag{3.150}$$

$$\mathbf{J} = \mathbf{A}(\mathbf{r})T(\mathbf{r},t) - \frac{1}{2}\overline{\mathbf{B}}\nabla T(\mathbf{r},t). \tag{3.151}$$

For concreteness, we think of $\mathbf{r}$ as a displacement vector. The following derivation does not depend on this choice, while the physical interpretation of $\mathbf{A}$ and $\overline{\mathbf{B}}$ does of course depend on the nature of configuration space. Equation (3.150) is a *continuity equation*, linking the probability flux $\mathbf{J}$ to the time change of the probability density. It has no physical content other than conservation of probability. Equation (3.151) is a linear *constitutive equation*, linking the flux to the probability distribution and its gradient, through the vector function $\mathbf{A}$ and the tensor $\overline{\mathbf{B}}$, of which nothing is known *a priori*. These quantities express the physics of the problem at hand. The FP equation is defined in a spatial domain Ω, and must be completed by model dependent boundary conditions, e.g. by reflecting boundary conditions.

Unlike other types of master equations, FP equations afford the possibility to estimate $\mathbf{A}$ and $\overline{\mathbf{B}}$ rather directly using experimental data. To see how this can be done, consider the time development of T during a small time interval $(0, \delta t)$. With a sharp initial condition, T will remain nearly equal to zero except within a region which is centred at $\mathbf{r}_0$ and small in extension compared to Ω. The (possible) finiteness of Ω can then be neglected, and boundary conditions at infinity can be utilized. We assume that the probability flux $\mathbf{J}$ and the probability density T both vanish at infinity.

Under the above conditions, an approximate equation for the average displacement $E_{\mathbf{r}}(\delta t)$ is obtained multiplying Eq. (3.150) by $\mathbf{r}$, and then integrating over all space.[3] Approximating the time

[3]We apologize for any possible confusion arising from the fact that $E_{\mathbf{r}}(\delta t)$ is *not* a function of $\mathbf{r}$. The subscript $\mathbf{r}$ is only included to indicate the type of expectation value.

derivative by a finite difference and integrating by parts, we obtain the equation

$$\frac{E_{\mathbf{r}}(\delta t) - \mathbf{r}_0}{\delta t} = \int_{-\infty}^{\infty}\int_{-\infty}^{\infty}\int_{-\infty}^{\infty} \mathbf{J}(\mathbf{r}, \delta t)dxdydz \qquad (3.152)$$

where x, y and z are the Cartesian components of $\mathbf{r}$. Using Eq. (3.151), we now obtain

$$\frac{E_{\mathbf{r}}(\delta t) - \mathbf{r}_0}{\delta t} = \mathbf{A}(\mathbf{r}_0) + \mathcal{O}(\delta t), \qquad (3.153)$$

or, equivalently,

$$E_{\mathbf{r}}(\delta t) - \mathbf{r}_0 = \mathbf{A}(\mathbf{r}_0)\delta t + \mathcal{O}(\delta t^2). \qquad (3.154)$$

If we imagine sampling the evolution of the system stroboscopically every δt seconds, the l.h.s. of the equation is the average size of a 'jump', or change, in the displacement vector $\mathbf{r}$ over the short interval δt. Hence the name *jump moment* is often used for $\mathbf{A}$. Stroboscopic sampling of trajectories provides experimental information on the form of $\mathbf{A}$.

Exercise 45: Prove Eq. (3.152) using integration by part and the vanishing of the flux at infinity.

Exercise 46: Prove Eq. (3.154) using that T is very peaked at $\mathbf{r}_0$.

To see how the tensor $\overline{\mathbf{B}}$ is related to the covariance of the displacement, a few steps are required. Consider first the diagonal element $E(x^2) - E^2(x)$ of the covariance tensor. Using Eq. (3.154), we note that

$$E_x^2(\delta t) = x_0^2 + 2x_0 A_1(\mathbf{r}_0)\delta t + \mathcal{O}(\delta t^2). \qquad (3.155)$$

Retracing the steps already used to derive $E_{\mathbf{r}}(\delta t)$, we also find

$$\begin{aligned}\frac{E_{x^2}(\delta t) - x_0^2}{\delta t} &= 2\int_{-\infty}^{\infty}\int_{-\infty}^{\infty}\int_{-\infty}^{\infty} x\mathrm{J}_1(\mathbf{r}, \delta t)dxdydz \\ &= (2x_0 A_1(\mathbf{r}_0) + B_{11}) + \mathcal{O}(\delta t). \qquad (3.156)\end{aligned}$$

From the last two equations, we conclude that the diagonal elements of $\overline{\mathbf{B}}$ are

$$B_{11} = \frac{\text{Var}_x(\delta t)}{\delta t}, \quad B_{22} = \frac{\text{Var}_y(\delta t)}{\delta t}, \quad \text{and} \quad B_{33} = \frac{\text{Var}_z(\delta t)}{\delta t}. \tag{3.157}$$

Turning to the off-diagonal elements, we consider the average $E_{xy}(\delta t)$ of the product of the first two coordinates of the displacement vector. A calculation yields

$$\frac{E_{xy}(\delta t) - x_0 y_0}{\delta t} = \int_{-\infty}^{\infty}\int_{-\infty}^{\infty}\int_{-\infty}^{\infty} (yJ_1(\mathbf{r}, \delta t) + xJ_2(\mathbf{r}, \delta t) dx dy dz. \tag{3.158}$$

From Eq. (3.154) we obtain

$$E_x(\delta t)E_y(\delta t) = x_0 y_0 + \left(x_0 A_2(\mathbf{r}_0) + y_0 A_1(\mathbf{r}_0) + \frac{B_{12} + B_{21}}{2}\right)\delta t + \mathcal{O}(\delta t^2). \tag{3.159}$$

We thus find

$$\text{Cov}(x, y) = E_{xy}(\delta t) - E_x(\delta t)E_y(\delta t) = \left(\frac{B_{12} + B_{21}}{2}\right)\Big)\delta t + \mathcal{O}(\delta t^2). \tag{3.160}$$

By extension, the off-diagonal elements of the symmetric covariance matrix are all related to the averages of pairs of off-diagonal elements B_{ij} and B_{ji}, where i and j are $1, 2$ or 3. Noting that covariance data are the experimental input used to determine $\overline{\mathbf{B}}$, we can as well define $\overline{\mathbf{B}}$ to be a symmetric tensor, which is then proportional to the covariance matrix of the displacement vector. This relation can be written as the tensor equation

$$E_{\mathbf{r}\mathbf{r}^T}(\delta t) - E_{\mathbf{r}}(\delta t)E_{\mathbf{r}^T}(\delta t) = \overline{\mathbf{B}}\delta t, \tag{3.161}$$

where the superscript T indicates transposition. Since $\overline{\mathbf{B}}$ is symmetric and positive definite, it is invertible and can be diagonalized by an orthogonal transformation in $\mathcal{R}^3$.

3.8.1. Stationary Behaviour

In a stationary solution to Eq. (3.150) and (3.151), $\mathbf{J}$ is time and space independent. As $\mathbf{J}$ also vanishes at the boundaries it must then vanish everywhere. A *putative* stationary solution will have the form

$$P_{1,\mathrm{eq}}(\mathbf{r}) \propto \exp\left(2\int_0^{\mathbf{r}} \overline{\mathbf{B}}^{-1}\mathbf{A}(\mathbf{v})\cdot d\mathbf{v}\right), \tag{3.162}$$

where $\overline{\mathbf{B}}^{-1}$ is the inverse of $\overline{\mathbf{B}}$. Unless the integrand appearing in the expression is the gradient of a scalar function, i.e. unless

$$\overline{\mathbf{B}}^{-1}\mathbf{A}(\mathbf{v}) = -\nabla S(\mathbf{v}), \tag{3.163}$$

the line integral is path dependent, and the equation hence meaningless. Accordingly, we assume Eq. (3.163) and rewrite Eq. (3.162) as

$$P_{1,\mathrm{eq}}(\mathbf{r}) \propto \exp\left(-2S(\mathbf{r})\right). \tag{3.164}$$

To ensure the integrability of $P_{1,\mathrm{eq}}$ in an infinite domain, $S(\mathbf{r})$ must be positive and divergent for large values of $\mathbf{r}$.

Example 5: Let $\mathbf{v}$ be the velocity of the Langevin particle moving in a large liquid filled vessel, which we treat as being infinitely large (see Eq. (3.1)). The liquid is isotropic, and the tensor $\overline{\mathbf{B}}$ is hence also isotropic, i.e. $\overline{\mathbf{B}} = B\overline{\mathbf{I}}$ where B is a scalar. Note that there are no explicit dependences on position in this problem, i.e. the configuration space is a 'velocity space'. The force on the particle is $-\gamma\mathbf{v}$, hence

$$\mathbf{A}(\mathbf{v}) = -\frac{\gamma}{2m}\nabla(\mathbf{v}\cdot\mathbf{v}). \tag{3.165}$$

The equilibrium distribution is Gaussian,

$$P_{1,\mathrm{eq}}(\mathbf{r}) \propto \exp\left(-\frac{\gamma\mathbf{v}\cdot\mathbf{v}}{mB}\right). \tag{3.166}$$

Comparing with the equilibrium distribution from statistical mechanics,

$$P_{1,\text{eq}}(\mathbf{r}) \propto \exp\left(-\frac{m\mathbf{v}\cdot\mathbf{v}}{2k_BT}\right)$$

we find

$$B = \frac{2k_BT}{m}\frac{1}{\tau}, \tag{3.167}$$

where $\tau = m/\gamma$, see Eq. (3.2). Using Eq. (3.157) the variance of the velocity can be determined experimentally. This constrains the values of the physical constants appearing in Eq. (3.167).

Example 6: Consider a particle diffusing freely in physical space. In the absence of forces, $\mathbf{A} = 0$. Hence no stationary solution exists which satisfies the necessary integrability conditions in an infinite domain. In a finite domain, the uniform distribution is integrable. In any case, we can still use Eq. (3.157) to identify $\overline{\mathbf{B}}$ as *twice* the *diffusion tensor*:

$$\overline{\mathbf{B}} = 2\overline{\mathbf{D}}. \tag{3.168}$$

In many applications, e.g. diffusion in liquids or gases, $\overline{\mathbf{D}}$ is a scalar. In contrast, the motion of defects in anisotropic crystals is described by an anisotropic diffusion tensor.

Example 7: The Smoluchowski equation describes the motion of a particle in a space varying force field. The equation only applies in the fully damped limit, where force and velocity are proportional. Assume that space is isotropic, and that the force can be written as

$$\mathbf{F}(\mathbf{r}) = -\nabla E(\mathbf{r}), \tag{3.169}$$

where $E(\mathbf{r})$ is the potential energy. Since $\mathbf{A} \propto \mathbf{F}$, identifying Eq. (3.164) with the Boltzmann distribution, which is proportional to $\exp(-\mathbf{E}(\mathbf{r})/k_BT)$, leads to

$$\mathbf{A}(\mathbf{r}) = -\frac{B}{2}\frac{\nabla E(\mathbf{r})}{k_BT}. \tag{3.170}$$

The Smoluchowski equation can thus be written as

$$\frac{\partial T(\mathbf{r},t)}{\partial t} = D\nabla \cdot \left[\frac{\nabla E(\mathbf{r})}{k_B T} T(\mathbf{r},t) + \nabla T(\mathbf{r},t)\right], \tag{3.171}$$

where D is the (isotropic) diffusion coefficient.

Analytical solutions of time dependent FP equations are possible in special cases, e.g. when boundaries have simple geometries and when **A** is a linear function. The eigenvalue expansions discussed previously do apply. To some extent, eigenvalues and eigenvectors can be obtained numerically, by discretizing the FP equations.

4

Monte Carlo Methods

4.1. INTRODUCTION

Simulations with a stochastic element are ideal for numerical experimentation with complex systems. Among these, Monte Carlo (MC) methods utilize random numbers to estimate probabilities, expectation values and other quantities of interest. Consider, e.g., a random walk on the real line. The position of the walker at (discrete) time t is the sum of t steps, or displacements, performed by the walker since starting. These steps are assumed to be statistically independent and identically distributed random variables drawn from a given distribution, e.g. a standard Gaussian. Generating these displacements and adding them up produces one trajectory of the random walker. Repeating the simulation M times allows the estimation of, e.g., the standard deviation of the position of the walker as a function of t. In so far the random walk is a model of a physical process,[1] the procedure above uses a Monte Carlo technique to calculate the properties of that model.

The calculation of the area enclosed in a curve inscribed in a unit square is a different application of MC techniques: Pairs of real numbers (x_i, y_i), for $i = 1, 2, \ldots M$ are drawn independently from a

[1]Random walks have been used as models of a plenitude of phenomena in, say, physics, economics and biology. A famous example is Einstein's investigation of Brownian motion.

uniform distribution in the unit interval, each pair identifying a point in the unit square. If the point lies inside the curve, the corresponding draw is a 'hit', otherwise it is a 'miss'. Since the probability of a hit equals the area of the curve, the latter can be estimated as the ratio of the number of hits to the total number of draws. This method is computationally wasteful in the low dimensional case just discussed, but works fine in analogous high dimensional spaces. An interesting problem arises if the area enclosed in the curve happens to be very small compared to the area of the unit square. If, e.g., the ratio is 10^{-20}, of the order of 10^{20} trials are needed to get a single hit, and the estimation task becomes lengthy and inaccurate on the computer.

In general, MC methods are computationally efficient for calculating definite integrals in high dimensional spaces and indispensable to estimate the equilibrium properties of models of physical systems. Kinetic Monte Carlo Methods (KMC) additionally provide a realistic description of the stochastic dynamics of complex systems.

This chapter deals with basic MC techniques based on Markov chains. These techniques are useful for modelling purposes, provide an application of the theoretical concepts previously discussed in this book and serve as an introduction to a vast area of computational science. The reader is referred to the extensive literature for a more in-depth study of the subject, e.g. Ref. [BH97, NB99].

4.2. IMPORTANCE SAMPLING

The seminal paper of Metropolis, Rosenbluth, Rosenbluth, Teller and Teller "Equation of State Calculations by Fast Computing Machines" [MRR+53], has now been cited more than 10000 times. The paper's popularity reflects the robustness and versatility of the so-called *Metropolis algorithm* it introduces.

To set the stage for a discussion of the algorithm, consider a discrete configuration space S containing a finite (but large) number $M = |S|$ of elements. The space can arise from the tessellation of a bounded region of a continuous configuration space, i.e. the positions of a set of particles confined in a box. It can also be directly part of a model, e.g. the N-dimensional hypercube containing all possible states of N Ising spins.

The properties of the model system considered are usually described by a set of real functions defined on the space $\mathcal{S}$, e.g. for concreteness a probability distribution P and a function f. Quantities to be estimated by MC methods are expectation values, i.e. sums of microstates of the general form

$$\mu_f = \sum_{i=1}^{M} f(i)P(i). \tag{4.1}$$

An exact evaluation of the above sum would require calculating for each state the values of P and f. This is computationally inconvenient if the cardinality M of the space is large. A typical example from physics would feature an Avogadro's number $N_A \approx 6{\cdot}10^{26}$ of particles each having two states, leading to $M = |\mathcal{S}| = 2^{N_A}$. In biology or sociology one may be dealing with fewer components than N_A, but the total number of configurations can still be enormous since it is given by the number of possible states of an individual component raised to a power N equal to the number of components.

We conclude that a direct and complete summation in Eq. (4.1) is impossible, which means that the sum must be estimated by sampling a smaller number of terms. This can be done because the vast majority of the terms contribute nearly zero to the sum, when, as often happens, the probability distribution P is sharply peaked on a small subset of 'important' configurations. The terms included in the estimation must therefore be sampled by drawing them from a distribution, G, which is strongly biased towards the same important contributions as P is, and which, in this sense, is 'close' to P. This type of approach is generally known as *importance sampling*. Using importance sampling, the sought expectation value is estimated as

$$\mu_f \approx \frac{1}{N}\sum_{i=1}^{N} f(i)\frac{P(i)}{G(i)}, \tag{4.2}$$

where N is a large integer. The challenge in importance sampling is to find a distribution G which is close enough to P and which is easy to handle on the computer.

Different Monte Carlo techniques based on importance sampling are used to extract physical parameters from model simulations. The example below deals with the estimation of the critical temperature of a model system which undergoes a second order phase transition. A transition at temperature $T = T_c$ is characterized by an *order parameter* which is equal to zero in the disordered phase, $T > T_c$, and different from zero in the ordered phase, $T < T_c$. Formally, phase transitions and their properties are well defined in the limit of infinite system size, the so-called thermodynamic limit. Unfortunately, a computer can only deal with systems of finite size. Using Binder cumulants [BH92] it is possible to estimate the critical temperature using data from finite systems.

Example 1: Binder cumulants.
For concreteness, we consider a version of the Ising model introduced in example (3.3). The spins $\sigma_1, \sigma_2, \ldots \sigma_N$ are located on a cubic d-dimensional lattice of linear size L. Each spin i is coupled through a ferromagnetic (i.e. positive) coupling to its neighbour spins. We denote the set spins consisting of the lattice neighbours of spin number i by $\{\mathcal{N}(i)\}$. And define the energy of a configuration by

$$H(\sigma_1, \sigma_2, \ldots \sigma_N) = -\frac{1}{2}\sum_{i=1}^{N} \sum_{j\in\mathcal{N}(i)} \sigma_i \sigma_j.$$

In this case, a suitable order parameter is the average of the magnetization

$$M = \sum_{i=1}^{N} \sigma_i, \quad N = L^d. \tag{4.3}$$

For $d \geq 2$, the magnetization performs stationary fluctuations around a zero average for $T > T_c$, and around a non-zero average for $T < T_c$. Hence, the form of the distribution of M changes dramatically near T_c. To see how that happens, consider the *spatial correlation function* of spins located k lattice units apart (in any direction). A successful and widely used approximation, *finite size scaling,* posits that $\xi(T)$ is the only *physical* length scale of importance for the dynamics, the other length scale being the linear size of the system L.

Denoting expectation values by the symbol E, the spatial correlation is then given by

$$\mathrm{Corr}(k) = E(\sigma_0 \sigma_k) \approx e^{-k/\xi(T)}.$$

At high temperatures, $T \gg T_c$, the correlation length is small, $\xi \ll L$, and the system consists of $(L/\xi)^d$ patches, or domains, each approximately comprising ξ^d correlated spins. These domains contribute additively and independently to the magnetization M of the full system. The probability density function (PDF) of M is hence well approximated by a Gaussian distribution with variance $\sigma_M^2 \propto (L/\xi)^d$.

The Binder cumulant is given in terms of the second and fourth moment of the magnetization by the expression

$$U_B(L,T) = 1 - \frac{E(M^4)}{3E^2(M^2)}. \tag{4.4}$$

Its value vanishes for a Gaussian distribution. The Binder cumulant hence generally gauges the deviations of the PDF of the magnetization from a Gaussian PDF.

For $T \to 0$, the Ising system approaches a ground state, where all spins are uniformly either 'up' or 'down'. Hence $U_B(L,0) = 2/3$. For $T \to \infty$, the distribution is, as mentioned, Gaussian and hence $U_B(L,\infty) = 0$. At intermediate temperatures, the cumulant scales with the number of domains, i.e. with the ratio $L/\xi(T)$.

In an infinite system, $\xi(T)$ diverges for $T \to T_c$. However, in a finite system the growth of the correlation length must saturate at the highest possible value, $\xi = L$. Clearly then, $U_B(L,T)$ is *independent* of L at the critical temperature. Plotting $U_B(L,T)$ versus T for several values of L identifies T_c as the temperature value at which the different Binder cumulants intersect. The characteristic features of Binder cumulants are illustrated in Fig. (4.1).

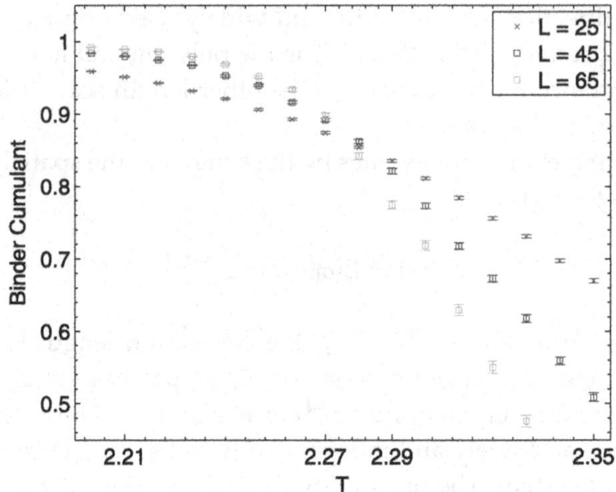

Figure 4.1. Binder cumulants, calculated using the Metropolis algorithm (see below) for three instances of the 2-d Ising ferromagnetic model. The linear sizes of the systems are shown in the legend. The cumulants are rescaled to approach unity when $T \to 0$. The three curves cross at a value close to the critical temperature of the model.

4.3. THE METROPOLIS ALGORITHM

The Metropolis algorithm solves the moment estimation problem using importance sampling, combined with the choice $G = P$ in Eq. (4.2). States are sampled according to a Markov chain $T(i, t|i_0)$ which, in the limit $t \to \infty$, relaxes to the distribution $P(i)$. The desired sampling is thus reached asymptotically for large values of t.

In general, a Markov chain is specified by a set of transition probabilities p_{ij} between states j and i. The first step in the construction of a Metropolis algorithm is hence to define a set of allowed transitions, i.e. a set of 'nearest neighbour' pairs of states for which the transition probabilities are not zero. These nearest neighbour transitions are the elementary moves of the Markov chain, and are collectively referred to as the *move class*. The choice of move class endows the state space with the structure of a graph, where the states are the nodes, and where the edges connect neighbouring states. The graph must be *i)* regular, i.e. all nodes must have the same number of edges impinging on them, and *ii)* connected, i.e. any two nodes must be linked by a sequence of edges. This last property is called *ergodicity*. Under such conditions, the stationary distribution of the Markov chain is independent of the move class, as required. However, it is important to note that the move class may strongly affect the rate at which the stationary distribution is approached. Indeed, special move classes have been devised to speed up convergence in specific situations, i.e. when the correlation time of the Markov chain diverges near a second order phase transition.

Assuming that the distribution P is nowhere equal to zero, we define the 'energy' function $\epsilon = -\ln(P)$. In physical applications, P is usually the Boltzmann equilibrium distribution of statistical mechanics. The function ϵ is then, modulo a constant, the energy divided by the temperature. Irrespective of the interpretation, the non-zero transition probabilities of the Markov chain generated by the Metropolis algorithm are

$$p_{ij} = \min(1, e^{\epsilon(j)-\epsilon(i)}). \tag{4.5}$$

These probabilities obey detailed balance with respect to the stationary distribution P, i.e.

$$P(i)p_{ij} = p_{ij}P(j). \tag{4.6}$$

The detailed balance together with the arguments given on p. 59 ensure that the Markov chain converges to P, as required.

When ϵ is related to the energy, the essence of Eq. (4.5) can be metaphorically expressed as follows: 'Downhill' transitions (those which decrease the energy) are always accepted, while 'uphill' transitions are accepted with a probability equal to the exponential of the energy difference between the initial and final states, divided by the temperature.

In a simulation, the state j is given as the 'current' state of the system and a prescription is needed to identify, among all other possible states, the 'target' state i. This prescription, which obviously defines the nature of the transitions allowed, can be specified in multiple ways.

Example 2: The N-dimensional hypercube is the configuration space for the Ising spin system. One can define two states to be neighbours if and only if they differ in the value of a single bit. Obviously, each state has N neighbours, and any state can be reached from any other using at most N elementary moves. A simple choice specifies the target state as one of the neighbour states on the hypercube, i.e. a transition will amount to picking at random and with equal probability one of the N spins as a candidate for a flip. The graph of this Markov chain is then regular and connected.

Example 3: A subset of the N-dimensional hypercube defines the configuration space of a *lattice gas*: Positions $1, 2, \dots N$ on a lattice in real space are either empty or occupied by a single particle. A configuration is a string of N bits, with values 0 and 1 representing an empty and a filled state, respectively. The simplest choice of move involves in this case a *pair* of sites, with a particle switching from one site to the other. We assume for simplicity that the choice of pairs is unrestricted, i.e. that a particle can jump within the lattice from any filled to any empty site. Since this move class conserves the number of particles, the configuration space of the problem is the subset of the hypercube containing the initial state and all states connected to it by particle swaps.

Example 4: The Ising model provides a simple and highly successful description of para- and ferro-magnetism. It has further applications, e.g. to social interactions. The model considers N spins, i.e. binary variables with values $\sigma_k = \pm 1, 1 \leq k \leq N$. These variables can be arranged, say, on regular grids in Euclidean space in d spatial dimensions or even on arbitrary networks defined by a connectivity matrix. Independently of real space properties, the model's configuration space consists of the 2^N vertices of the N-dimensional hypercube. Each state has precisely N neighbours, each differing in the value of one variable, i.e. a single bit.

The first part of the Metropolis algorithm consists in picking a neighbour 'at random', i.e. one of the N neighbours of the current state is chosen with probability $1/N$. The acceptance or rejection of the candidate move thus specified requires evaluating the interaction energy. The latter has various forms, depending on the model under consideration. When modelling ferromagnetic behaviour, the energy of spin j is given by

$$e(j) = -\sigma_j \sum_{k \in \mathcal{N}_j} \sigma_k, \tag{4.7}$$

where $\mathcal{N}_j$ is the neighbourhood in *real space* of spin j, e.g. if the spins are arranged on a 2-d square lattice, each spin has four neighbours, located one lattice site away. Equation (4.7) expresses the propensity of neighbouring spins to point in the same direction, since a parallel alignment of the spins gives a lower energy. The energy of a configuration is

$$E = -\frac{1}{2} \sum_{j=1}^{N} e(j), \tag{4.8}$$

where the factor $\frac{1}{2}$ offsets that the contribution from each spin pair is counted twice in the sum.

Example 5: The Metropolis algorithm can also tackle problems where configuration space is continuous. Such spaces are often called *energy landscapes*. We envisage N interacting 'particles', e.g. molecules, or individuals, each moving in a three-dimensional Euclidean space. Let the tuple $\mathbf{r}_j = (r_{j,x}, r_{j,y}, r_{j,z})$ represent the coordinates of particle j. For each point of the landscape, $\mathbf{r}_1, \mathbf{r}_2, \ldots \mathbf{r}_M$, we define the energy as

$$E(\mathbf{r}) = \frac{1}{2} \sum_{ij} f(|\mathbf{r}_i - \mathbf{r}_j|), \tag{4.9}$$

where **f** defines a distance dependent interaction energy between pairs of particles. A candidate configuration is generated from the current configuration in two steps. First a particle is selected with probability $1/N$ among the N available choices. Second, a candidate position for that particle is obtained by adding to its current position a random *increment*. Typically, this increment has a multivariate Gaussian distribution with zero average. Finally, the candidate position is either accepted or rejected according to the Metropolis rule.

In Monte Carlo algorithms designed to sample equilibrium properties, the unit of time is of scarce relevance, and, in fact, remains formally undefined. Whenever non-equilibrium situations are considered, it is of interest to relate the Monte Carlo steps to time. Even though this is a non-trivial problem, a practical way to do so is to introduce a discrete time whose unit, which is called a *Monte Carlo sweep*, consists in a system with N components of N consecutive MC steps. On average, each component is then queried once during one Monte Carlo sweep.

The Metropolis algorithm generates a trajectory in configuration space, from which (Monte Carlo) time series $f(1), f(2), \ldots f(t)$, where f is any function of interest, can be calculated. In general, such time series will be *non-stationary* and *correlated*. These properties negatively affect the accuracy of the estimated equilibrium average μ_f given by Eq. (4.2). To ensure a good accuracy, the length t of the time series must be a *large* multiplum of the correlation time τ_f of the series. This aim is particularly time consuming to reach when the correlation time τ_f is large, e.g. near second order phase transitions.

Associated with a divergent correlation length is a divergent correlation time. To reduce the computational time near a second order phase transition, more refined move classes are available for Ising systems. In the Wolff and Swendsen–Wang algorithms whole clusters of spins with the same orientation are flipped together. The clusters are constructed probabilistically, starting from a randomly chosen spin and adding spins, one at the time with a given probability P_{add}. By tuning P_{add}, it is possible to achieve a high speed of convergence to equilibrium. However, the dynamics of such algorithms lacks a physical interpretation.

Convergence to the required solution can also be excruciatingly slow, but for different reasons, in low temperature MC simulations of systems with many local energy minima: Once a trajectory hits a local energy minimum, all transitions are 'uphill' and many attempted moves are typically rejected before a new configuration can be sampled, which is wasteful of computer time. The difficulty often arises in, e.g., optimization problems, where the ground state configuration of a given model system is sought by running a MC simulation in which the temperature is slowly decreased, a process known as *simulated annealing* [SSF02]. A family of MC techniques called *parallel tempering* or *replica exchange methods* is very successful in circumventing this type of problem. In parallel tempering several copies, or replicas, of the same system are updated independently and in parallel. Each of the N replicas is run at a different temperature T_i, $1 \leq i \leq N$. There are two different types of moves. The first type uses the standard Metropolis acceptance criterion. Consider, e.g., the replica run at temperature T_k. A candidate configuration j is generated from one of the neighbours of the 'current' configuration i. Denoting the energy difference between the two by $\delta_{ij} = E_i - E_j$, and explicitly introducing the temperature in Eq. (4.5), a move to the new configuration is accepted with probability

$$p_{ji,T_k} = \min(1, e^{\delta_{ij}/T_k}). \tag{4.10}$$

The second type of move is a switch of configurations between replicas run at temperatures T_k and T_l, i.e. the replica run at T_k goes from its current configuration i to the current configuration j of the replica run at T_l, while the opposite happens for the latter. For each of the replicas, the move would be accepted with a probability given by Eq. (4.10). Simultaneously accepting both moves has therefore probability $p_{ji,T_k,T_l} = p_{ji,T_k} \cdot p_{ij,T_l}$, i.e.

$$p_{ji,T_k,T_l} = \min(1, e^{\delta_{ij}(1/T_k - 1/T_l)}). \tag{4.11}$$

Qualitatively, parallel tempering works by enlarging the class of candidate moves at one's disposal. If the two replicas involved in a switch are run at vastly different temperatures, the energy difference between the two configurations to be exchanged will typically be numerically large, and the r.h.s. of Eq. (4.11) will correspondingly be small. In general, a successful implementation of parallel tempering requires careful consideration of several issues related to the problem at hand and to the computer resources at one's disposal, e.g. how many different replicas should be run and at which temperatures, how often should configuration switches be attempted relative to 'standard' moves and should these switches only involve 'neighbouring' temperatures or not.

We finally turn to methods now variously called rejectionless, *event driven* or Kinetic Monte Carlo (KMC) methods, which, in contrast to those just discussed, have physically plausible dynamics.

4.4. EVENT DRIVEN ALGORITHMS

Convergence to equilibrium can be slow if a random sequential MC algorithm lingers near metastable configurations where rejections are highly probable. The process can be accelerated by considering, at each stage, a list of possible dynamical events and choosing one of them with the 'correct' probability, i.e. the probability leading to the desired equilibrium state. Apparently, the idea was introduced by J.-M. Lanore [Lan74] in connection with point defect diffusion in solids and soon thereafter developed by Bortz, Kalos and Lebowitz [BKL75] for Ising spin systems. Gillespie [Gil77] later applied the same general approach to chemical reaction dynamics. In all cases, the MC is accelerated by avoiding rejections. This is achieved at the price of increased memory requirements and coding complexity.

KMC methods address systems consisting, as usual, of N interacting components. Each possible move comprises the random selection of one component, followed by a deterministic update of the respective configuration which excludes rejections. The Markov chain does move at each step from one configuration to a different one. However, physical and model systems alike would usually not spend an equal time in all configurations. In order to correctly estimate expectation values from the trajectory of the Markov chain, a random 'residence' time must, for each state, be drawn from a suitable distribution. A different event driven algorithm is the Waiting Time Algorithm [DS01] (WTM), where stochastic waiting times for all possible moves are calculated and ordered, and where the move with the shortest waiting time is the one actually performed.

Beside asymptotically sampling a desired equilibrium distribution, KMC methods closely mimic the stochastic dynamics of multicomponent systems. This is of particular interest in models of complex systems, where the equilibrium distribution is de facto not achievable which makes dynamical properties all important.

Assume for concreteness that the system components are located in a 3D 'physical' space. Each component, $i = 1, 2, \ldots N$, is endowed with a set of 'inner' states, indexed by Latin letters, e.g. the inner states of an Ising spin are $a = \pm 1$. For molecules 'a' is a positional label, spanning over a set of integers. In all cases, the energy $E_i(a, \{b_k\})$ of component i depends on the state a of the component itself and on the states b_k, $k = 1, \ldots z_i$ of its z_i neighbours in *real space*. The number of neighbours can either be fixed, or it can vary dynamically, e.g. each spin placed on a 3D cubic lattice has invariably six neighbours, while the number of neighbours of a particle in a chemical process can vary over time. The rate at which a component changes its state from a to a' in a single update, is taken to be

$$r_i(a', a) = \exp\left(-\frac{\delta E_i(a', a)}{2T}\right), \quad 1 \leq i \leq N, \tag{4.12}$$

where $\delta E_i(a', a) \stackrel{\text{def}}{=} E_i(a', \{b_k\}) - E_i(a, \{b_k\})$, and where T is the temperature. The factor of 2 in the denominator ensures that the ratio of the rates fulfils the detailed balance condition.

To simplify the notation, we arrange all possible transition rates into a one-dimensional array r_k, where $k = 1, 2 \ldots k_{\text{max}}$. Each k corresponds to a specific component, and specific initial and final states.

As it will become apparent, event driven algorithms involve the searching and/or reordering of large databases at each update. The computational load introduced by these operations is considerable, and may well offset the time-savings obtained by avoiding rejections. For an efficient implementation of event driven algorithms adequate searching and manipulation techniques are indispensable. The discussion below only focusses on the general structure and physical implications of the methods. Any technical questions of implementation are left to the specialized literature.

4.4.1. Standard KMC Methods

A standard KMC comprises the following steps:

1. Pick an initial configuration, specified by the state a_i of each particle i.
2. Set the time to $t = 0$, and the loop index to $l = 0$.
3. Order all transitions rates in the one-dimensional array r_k, $k = 1, 2 \ldots k_{\max}$.
4. Calculate the cumulative probability distribution

$$U(k) = \left(\sum_{i=1}^{k_{\max}} r_i \right)^{-1} \sum_{i=1}^{k} r_i. \tag{4.13}$$

   Note that the constant prefactor is included to ensure the correct normalization.
5. With probability $U(k) - U(k-1) \propto r_k$, pick transition k and change the state of the system accordingly.
6. Draw a time increment Δt_l from the exponential distribution with average $1/r_k$.
7. Increment the time: $t \to t + \Delta t_l$, and update the loop index: $l \to l + 1$.
8. If $t > t_{\text{stop}}$ or $l > l_{\text{stop}}$, stop the simulation. Otherwise, go to 3.

Ergodicity hinges on the definition of the allowed transitions, and can only be discussed in a specific setting. By Eq. (4.12), the ratio of a transition rate to the rate of the reverse transition satisfies detailed balance. The Markov chain hence converges to the desired equilibrium distribution.

Estimated expectation values have the form

$$\mu_f = \frac{1}{T} \sum_{i=1}^{N} \Delta t_i f(i), \quad T = \sum_{i=1}^{N} \Delta t_i, \tag{4.14}$$

where the weight Δt_i accounts for the time the system would dwell in configuration i, if updated according to the standard Metropolis algorithm. The microscopic time unit is implicitely set to one in calculating the Δt_i. However, Eq. (4.14) is clearly not affected by this choice.

While the Δt_i are needed for estimating expectation values, they do not represent an ordering of the dynamical changes in time. This means that the KMC method cannot be used to simulate the temporal evolution of the system. This is in contrast to the Waiting Time Method which does generate a representation of the dynamics. We turn to this method in the next section.

4.4.2. Waiting Time Method

In a multicomponent physical system, changes occurring somewhere *during* the time Δt_k, can affect the interaction patterns and hence the transition probabilities. This aspect is taken into account in the Waiting Time Method, where the component with the *shortest waiting time* is always updated first. The waiting times are here the only random variables entering the dynamics.

The Waiting Time algorithm comprises the following steps:

1. Pick an initial configuration, specified by the state a_i for each particle i.
2. Set the 'global time', time to $t_{\text{global}} = 0$, and the loop index to $l = 0$.
3. List all possible transitions rates as a one-dimensional array r_k, $k = 1, 2 \ldots k_{\max}$.
4. For each k, draw a random number u_k from the uniform distribution in the unit interval. Form a list of waiting times having elements

$$\tau_k = -\ln(u_k)/r_k, \tag{4.15}$$

where r_k is given by Eq. (4.12).
As defined above, the waiting times are independent and exponentially distributed random numbers, in accordance with the idea that changes of state are, for each component, a Poisson process whose average only depends on the local interactions.

5. Set $t_{\text{next},k} = t_{\text{global}} + \tau_k$. Transition k would occur at time $t_{\text{next},k}$ in a fixed environment. Note that the environment remains fixed until the transition with the shortest $t_{\text{next},k}$ occurs.
6. Identify the smallest update time, $t_{\text{next}} = \min_k(t_{\text{next},k})$, and carry out the corresponding transition.
7. Increment the global time: $t_{\text{global}} = t_{\text{next}}$, and the loop index $l \to l+1$.
8. Recalculate the waiting times τ_k for the transitions affected by the last move, i.e. the transitions involving components whose local interactions have changed. Since the state changes are memoryless Poisson processes, there is no need to recalculate the other waiting times.
9. If $t > t_{\text{stop}}$ or $l > l_{\text{stop}}$, stop the simulation. Otherwise, go to 5.

As for the KMC methods previously discussed, ergodicity can only be dealt with in the context of a specific choice of allowed transitions. To show that the WTM asymptotically samples the Boltzmann equilibrium distribution, it suffices to consider an arbitrary infinitesimal time interval $(t, t + \delta t)$. Since all transitions are memory-less Poisson processes, we can without loss of generality choose $t = 0$ and use that transition k happens with probability $\exp(-r_k \delta t) r_k \delta t$. Since δt is small, we only need the corresponding first order approximation $r_k \delta t$. Due to the form of the transition rate, see Eq. (4.12), detailed balance is fulfilled, whence the desired result follows.

The 'global time' of the WTM is, for a sufficiently large number N of components, independent of N, but unfortunately, the computation time required to carry out a calculation up to a given global time is not independent of N: As N increases, the number of transitions in a fixed interval δt increases in proportion to N, mirroring the behaviour of a physical system. Secondly, if a move happens to create a highly unstable situation, where several components have negative energy barriers, these components acquire high transition probabilities and are, with high probability, updated within a short time. This endows the WTM with the ability to follow spatio-temporal events, e.g. clusters of spatially contiguous components changing state in a coherent fashion.

5

Record Statistics and Extremal Statistics

5.1. INTRODUCTION

As witnessed by the famous *Guinness Book of Records*, record events of all kinds have always attracted a great deal of interest. Aside from their popular appeal, extremes and their properties are important topics of mathematical statistics [Gli78, Kru07]. Furthermore, the behaviour of complex systems is often determined by the unusual rather than the average or typical events, and in some contexts record-sized events have important implications, e.g. in population dynamics mutants surpassing the current standard for highest reproductive success contribute disproportionately to the genetic pool of the next generation. Record events might therefore be of significance to biological evolution, as in particular suggested by the catchphrase 'survival of the fittest'.[1] Darwinism's impact on biology—and human culture at large—can hardly be exaggerated. In recent years computer optimization methods have appeared, e.g. Genetic Algorithms [Gol89] and Extremal Optimization [BP01], which are inspired by evolution dynamics and which, implicitely or explicitely, utilize records. In a computer program, or in a

[1]The phrase is attributed to Herbert Spencer in *The Principles of Biology*, 1864. As discussed in detail in Chapter 12, we find the underlying concept problematic.

social context, the function, entity or organization in charge of memorizing the existing record and detecting its eventual obliteration is clearly distinct from the process being monitored. By contrast, record keeping in biological evolution is fully integrated in the dynamics. In cases where record fluctuations act as seeds for an amplifying dynamical process, the magnitude of the seed bears no direct relation to the magnitude of the final outcome of the amplification, e.g. a snow avalanche can be triggered by sound vibrations above a certain threshold. However, its impact is more related to the amount of snow on the ground and to the slope and shape of the valley than to the intensity of the sound.

How record-sized events in a stochastic time series may be related to, and indeed be a determining factor for, the dynamics of complex systems is discussed in some detail in Chapters 9 to 13. In the following we focus on the temporal statistics of records: How records within a stationary and δ-correlated time series are statistically distributed in time. Secondly, we analyse records in a time series which is both non-stationary and correlated: The position of a random walker moving along a line. Finally, we consider the distinct issue of extremal value statistics, i.e. the distribution of the largest of n observations.

5.2. RECORDS IN WHITE NOISE

Record-sized entries within a stationary series of statistically independent and identically distributed random numbers, so-called *white noise*, form a subseries whose interesting statistical properties follow from simple arguments [Gli78, SL93, SBA98, Nev01], e.g. since it gets increasingly harder for an entry to qualify as the next record, records appear at a decreasing rate and the subseries of record-sized entries is clearly not stationary. When the stationary time series from which the records are extracted is *white noise* the number of records in the time interval $(1, t)$ is to a good approximation Poisson distributed with average $\ln t$.

Equivalently, if the k'th record appears at times t_k, the ratios $\ln(t_k/t_{k-1})$ are independent random variables with an exponential distribution.

Consider first an ordered sequence of $l+1$ random numbers drawn independently from a common distribution, and indexed from 0 to l. We exclude in the following distributions supported on a finite set, as these eventually lead to a record which can never be beaten. Apart from this modest constraint, the form of the distribution is immaterial. By definition, the first entry of the sequence ($k=0$) is a record. Other entries are records if and only if they are larger than all their predecessors.

Next, denote by $P_n(l,[0,m_1\dots m_n])$ the probability that n records are located at entries $0, m_1, \dots, m_n$, and by $p_n(l)$ the probability that n out of the $l+1$ entries are records, regardless of their location. Clearly, in the case $n=1$, the first entry must be the largest. As the largest entry can be anywhere with equal probability,

$$P_1(l,[0]) = p_1(l) = \frac{1}{l+1}, \quad l = 0,1\dots\infty. \tag{5.1}$$

For two records, one occurring at 0 and the other at $m_1 \leq l$, we find

$$P_2(l,[0,m_1]) = \frac{1}{m_1}\frac{1}{l+1}, \quad 1 \leq m_1 \leq l. \tag{5.2}$$

The first factor on the r.h.s. of the equation is the probability that the zero'th entry is the largest of the first m_1 entries, and the second is the probability that the largest entry of the whole series be located at m_1. Clearly,

$$p_2(l) = \frac{1}{l+1}\sum_{m_1=1}^{l}\frac{1}{m_1} = \frac{H_l}{l+1} \tag{5.3}$$

where the *harmonic number* $H_l = \sum_{k=1}^{l} 1/k$ satisfies

$$H_l = \ln(l+1) + \gamma + \mathcal{O}(l^{-1}) \tag{5.4}$$

with $\gamma = 0.57721\ldots$ being the Euler–Mascheroni constant. Turning now to a generic n, we find, by the same arguments,

$$P_n(l,[0,m_1,m_2,\ldots m_{n-1}]) = \frac{1}{l+1}\prod_{k=1}^{n-1}\frac{1}{m_k},$$

$$\text{with:}\quad 1 \le m_1;\quad m_{k-1} < m_k;\quad m_{n-1} \le l. \tag{5.5}$$

To obtain $p_n(l)$, the distribution $P_n(l,[0,m_1,m_2,\ldots m_{n-1}])$ must be summed over all possible values of $m_1, m_2 \ldots m_{n-1}$. Unfortunately, this rather cumbersome operation does not lead to a closed form expression. A closed form expression can however be reached in the continuum limit, where all sums turn into integrals. To this end, let δt be a small time interval, and assume that a time series is produced by sampling a stationary noise signal at regular intervals of duration δt. Furthermore, let $t_k \stackrel{\text{def}}{=} m_k\delta t$ and $t \stackrel{\text{def}}{=} l\delta t$ be the time of occurrence of the k'th record in the series, and the total observation time, respectively. Unless the noise is δ-correlated, contiguous values of the time series will of course become correlated in the limit $\delta t \to 0$. To avoid correlations, we now assume that the noise is white. Noting that $m_k = t_k/\delta t$, and taking the limit $\delta t \to 0$, we find

$$p_n(t) = \frac{1}{t}\int_{t_{n-2}}^{t}\frac{dt_{n-1}}{t_{n-1}}\prod_{k=2}^{n-2}\int_{t_{k-1}}^{t_{k+1}}\frac{dt_k}{t_k}\int_1^{t_2}\frac{dt_1}{t_1}. \tag{5.6}$$

As the integrand in the above expression is a symmetric function of the arguments $t_1,\ldots,t_{n-1}$, any permutation of the arguments does not change the integral. Summing over all permutations of the order of the $n-1$ variables and dividing by $(n-1)!$ leaves the expression unchanged and yields

$$p_n(t) = \frac{1}{t}\frac{1}{(n-1)!}\left(\int_1^t\frac{dz}{z}\right)^{n-1} = \frac{1}{t}\frac{\log(t)^{n-1}}{(n-1)!},\quad n = 1,2,\ldots\infty \tag{5.7}$$

The expression can be recognized as a Poisson distribution with expectation value

$$E(n) = \log(t). \tag{5.8}$$

Clearly, the number of records falling between times t_w and $t > t_w$ is the difference of two Poisson processes, and hence itself a Poisson process with expectation value $E(n) = \ln(t) - \ln(t_w) = \ln(t/t_w)$. The average number of records per unit of time decays as

$$\frac{dE(n)}{dt} = \frac{1}{t}. \tag{5.9}$$

A third property of Poisson distributions, namely the exponential distribution of the waiting time between successive events, leads in connection with Eq. (5.7) to the following: Denoting by $\tau_k = \ln t_k$ the logarithm of the time at which the k'th record occurs, the stochastic variables $\Delta_k = \tau_{k+1} - \tau_k = \ln(t_{k+1}/t_k)$, $k \geq 1$ and $\Delta_0 = \tau_1$ are statistically independent and exponentially distributed with unit average:

$$P_{\Delta_k}(x) = \text{Prob}(\Delta_k < x) = 1 - \exp(-x). \tag{5.10}$$

Since

$$\tau_k = \sum_{i=0}^{k-1} \Delta_i \tag{5.11}$$

is a sum of k independent exponential variables with unit average, it follows that τ_k is Gamma distributed, with density given by

$$P_{\tau_k}(t) = \frac{t^{k-1}}{(k-1)!} e^{-t}. \tag{5.12}$$

Both the average and the variance of the distribution are equal to k. By Jensen's inequality [Rud66] we find

$$\ln(E(t_k)) \geq E(\ln(t_k)) = k. \tag{5.13}$$

For large k, the Gamma distribution can be approximated by a Gaussian distribution, and the waiting time t_k is therefore approximately log-normal.

Exercise 1: Let t_{k-1} and $t_k > t_{k-1}$ be the times at which two successive records occur. For given t_{k-1}, define the conditional probability

$$P_{t_k|t_{k-1}}(x) = \text{Prob.}\left(\frac{t_k}{t_{k-1}} < x\right), \quad 1 < x < \infty. \tag{5.14}$$

Using Eq. (5.10), show that

$$P_{t_k|t_{k-1}}(x) = 1 - \frac{1}{x}, \quad 1 < x < \infty.$$

Show then that the conditional expectation value $E(t_k|t_{k-1})$ is infinite. Using similar steps, show that

$$E(t_{k-1}|t_k) = \frac{1}{2}.$$

Exercise 2: Princess Recordia of the kingdom of Math is soon to be married. Her M suitors, $M \gg 1$, are all invited to court. To speed up the selection, the Lord Chamberlain suggests that the suitors be introduced to Recordia one after the other, and that she decide, immediately after meeting a suitor, whether to marry him or to dismiss him and try the next one. The Queen Mother frets that, with such haste, the probability of choosing the best suitor would only be of order $1/M$. But the math savvy Recordia assures her that, using the right strategy, she can pick the best suitor with probability $1/e$. What is the strategy that Recordia has in mind?

Exercise 3: Using Eq. (5.13), show that, on average, the waiting time from the k'th to the $(k+1)$'th record grows at least exponentially with k.

A simple generalization of the results just obtained replaces the stationary time series by α mutually independent and stationary time series. For each of these, records are defined as discussed. The total number of records observed in a given time interval is then the sum of α independent Poisson variables, and hence itself a Poisson process. The expected number of records falling between times $t_1 < t$ and t is therefore

$$E(n) = \alpha \ln(t/t_1), \tag{5.15}$$

and the full distribution is

$$p_n([t_1, t]) = \left(\frac{t}{t_1}\right)^{-\alpha} \frac{(\alpha \ln(t/t_1))^{n-1}}{(n-1)!},$$
$$n = 1, 2, \ldots \infty; \quad \alpha = 1, 2, \ldots \infty. \tag{5.16}$$

To summarize, the number of records occuring between times t_1 and $t > t_1$ in a time series of stationary white noise is a Poisson process. The expectation value of the process is proportional to $\ln(t/t_1)$ *irrespective* of how the noise values are distributed. This result changes when the assumptions on the underlying time series are relaxed, see, e.g., Ref. [Kru07] for a discussion of non-stationary series. Below we consider a different case, where the series is correlated and non-stationary.

5.3. RECORDS AND FIRST PASSAGE IN BROWNIAN MOTION

The positions $x(t_0), x(t_1), \ldots$ of a random walker moving in an infinite one-dimensional lattice $\{\ldots - n, -n+1, \ldots 0 \ldots n, n+1, \ldots\}$ constitute a time series which is neither δ-correlated nor stationary. As we shall see, the statistical properties of the records in this type of time series are rather different from the case treated in the previous section.

Since the random walk problem is translationally invariant, we can assume, without loss of generality, that the walker is initially located at the origin, i.e. $x(t = 0) = 0$. As usual, the $t = 0$ entry counts as the first record, while subsequent records occur at times larger than their predecessors. The time T_1 elapsed between the first and the second record coincides with the time of *first passage* through position $x = 1$. In general, the time T_k elapsed from the k'th to the $k+1$'th record coincides with the time of first passage through position $x = k$ for a walker starting at $x = k - 1$. The T_k's are mutually independent and identically distributed stochastic variables. We can hence use the symbol T for the time of first passage to a site which is the right neighbour of the initial position.

The walker remains at its current position for a time T_r which is random and has an exponential distribution of unit average. Jumps are to the right or to the left with probabilities r_{r} and $1 - r_{\mathrm{r}} = r_{\mathrm{l}}$, respectively.

Let $f_{k,l}(t)$ be the probability density function (PDF) that the first passage through position k with start point at l occurs at time t. Since, as mentioned, this PDF only depends on $k - l$, we use $f_T(t)$ for any $f_{k,k-1}(t)$. Secondly, let $P_n(t)$ be the probability that precisely n records occur in the interval $[0, t)$. Clearly, $n > 1$ records in time t require $n - 1$ records at an earlier time $t' < t$. A single record at time t implies that the first passage through 1 is never achieved. In summary,

$$P_1(t) = 1 - \int_0^t f_T(t')dt' \tag{5.17}$$

$$P_n(t) = \int_0^t P_{n-1}(t')f_T(t - t')dt', \quad n = 2, 3, \ldots. \tag{5.18}$$

Similar arguments lead to an integral equation for f_T: The walker can jump to the right and hit site k at his first move. If the first jump is to the left, the walker has to return to his original position (zero) at time $t' < t$ before a renewed attempt to hit site 1 can be made. The corresponding PDF is denoted by $f_{k,k-2}$, and

$$f_T(t) = r_{\mathrm{r}}e^{-t} + r_{\mathrm{l}} \int_0^t e^{-t'} f_{k,k-2}(t - t')dt', \tag{5.19}$$

where r_{r} and r_{l} are the probabilities of jumping to the right and left, respectively. Furthermore, we note that

$$f_{k,k-2}(t) = \int_0^t f_T(t')f_T(t - t')dt' \tag{5.20}$$

is itself a convolution. All the equations above involve convolutions and can be solved using Laplace transforms which map convolutions into products. Using a tilde for Laplace transformed functions we obtain

$$\tilde{P}_1(s) = \frac{1 - \tilde{f}_T(s)}{s} \tag{5.21}$$

$$\tilde{P}_n(s) = \tilde{P}_{n-1}(s)\tilde{f}_T(s) \quad n = 2, 3, \dots, \tag{5.22}$$

and

$$\tilde{f}_T(s) = \frac{r_r}{(1+s)} + \frac{r_l}{(1+s)}\tilde{f}_T^2(s). \tag{5.23}$$

Eqs. (5.21) and (5.22) are solved by

$$\tilde{P}_n(s) = \frac{\tilde{f}_T^{n-1}(s) - \tilde{f}_T^n(s)}{s} \quad n = 1, 2, \dots. \tag{5.24}$$

Turning now to f_T, we note that the physically relevant solution of Eq. (5.23) must fulfil $0 < \tilde{f}_T(s) \leq 1$ for all s. We note that $\tilde{f}_T(s = 0) = \int_0^\infty f_T(t')dt'$ is the probability that the position $x = 1$ will be visited at all. As it turns out, for $r_l > r_r$ one has $\tilde{f}_T(s = 0) < 1$. The walker has in this case a non-zero probability of escaping to $-\infty$, without ever visiting $x = 1$. Correspondingly, there is a non-zero probability that the initial record will never be beaten. Intuitively, this is due to a net drift to the left. In the extreme case $r_l = 1$, the walker moves at constant negative velocity, and $x = 1$ can never be visited. For $r_r > r_l$ the solution of Eq. (5.23) is:

$$\tilde{f}_T(s) = \frac{1+s}{2r_l}\left(1 - \left(1 - \frac{4r_r r_l}{(1+s)^2}\right)^{1/2}\right), \quad r_r > r_l. \tag{5.25}$$

Exercise 4: Find the solution of Eq. (5.23) which holds for $r_r < r_l$. Determine how the probability that a second record never occurs depends on r_l.

Let us first consider the interesting case $r_l = r_r = 1/2$. There is no net average velocity, and, as we shall see, the time elapsed between successive records has infinite expectation value. A calculation gives

$$\tilde{f}_T(s) = \frac{2(1+s) - 2\,(s(2+s))^{1/2}}{2} = \frac{s^{1/2} - (2+s)^{1/2}}{s^{1/2} + (2+s)^{1/2}}. \tag{5.26}$$

Inverting the Laplace transform yields

$$f_T(t) = \frac{e^{-t}}{t} I_1(t), \tag{5.27}$$

where I_1 is the modified Bessel function of order one. Using a different route, the same result is derived by Feller, see Eq. (17.13) in Ref. [Fel66]. We note that

$$f_T(t) \to \frac{t^{-3/2}}{(2\pi)^{1/2}} \quad \text{for } t \to \infty. \tag{5.28}$$

Hence, f_T has no finite average, as anticipated. The Laplace transform of the expected number of record-breaking event can be calculated from Eq. (5.24) to be

$$\tilde{E}(n,s) = \frac{1}{s(1 - \tilde{f}_T(s))} = \frac{1}{s((2s)^{1/2} + \mathcal{O}(s))}. \tag{5.29}$$

The second equality is found by expanding $\tilde{f}_T(s)$ to lowest order in s. Inverting the Laplace transform of the r.h.s. of the above equation yields the asymptotic behaviour of $E(n,t)$ for large t and for $r_r < r_l$, i.e.

$$E(n,t) \approx (t/2)^{1/2}. \tag{5.30}$$

Records thus occur at a decreasing rate, as in the white noise case previously discussed. The important difference is that in the case of Brownian motion the form of the distribution of the single steps has a crucial role: When $r_r > r_l$, the walker moves to the right at the average speed $r_r - r_l$. Hence,

$$E(n,t) \approx (r_r - r_l)t. \tag{5.31}$$

5.4. EXTREMAL VALUE STATISTICS

Record and extreme value statistics both deal with large events, but the former is concerned with the *temporal* distribution of records in a time ordered series, while the latter is concerned with the *size* distribution of the largest element in a set of n random variables. Record statistics implies an ordering since a record refers to what has happened earlier: The world record for long jump was broken today because Mr. Tall managed to jump 4 mm longer than the previous record set by Mr. Short two years ago. Thousands of long jumps have been undertaken between those by Mr. Short and Mr. Tall. But none of these released a new record because they were all shorter than Mr. Short's. The focus of record statistics, at least in the context of our book and as discussed in the preceding sections, is to understand the statistics of the time instances at which the considered quantity achieves a value higher than any previous recorded value. One is less interested in the actual value of the record, here the length of the jump.

Addressing the long jump from the viewpoint of extreme value statistics, the question could, e.g., be the following. One hundred equally competent long jumpers gather to practice their passion for long jumps. We have been hired to construct the jump pit. Now the question arises: How long do we need to make the pit in order to make sure that the longest, i.e. the extreme, of the 100 jumps doesn't reach beyond the soft sand of the pit. Here we don't care about the sequential ordering of the 100 jumps, but only about the size of the extreme jump. Since if we ensure that the length of the pit is longer by a safety margin than the statistically expected extreme, all jumpers are safe. To do the statistics we obviously need information about the distribution of the length of the individual jumps.

Let us now develop the formalism used to treat such issues mathematically. Consider a sequence $x(t_i)$ of identically distributed and independent real valued stochastic variables for $t_i = 1, 2, 3, \ldots$. A record happens at time t_r if and only if

$$x(t_r) > x(t) \quad \text{for all } t < t_r. \tag{5.32}$$

As discussed in Section 5.2, the sequence of records falling between times t_1 and t_2 is to a good approximation a Poisson process whose average is proportional to $\ln(t_2/t_1)$. Remarkably, this result is largely independent of the distribution of $x(t)$. In contrast extreme value statistics is concerned with the functional form of the distribution of the greatest (or smallest) values of the x's. As one would expect, this distribution turns out to depend on the distribution of the individual variables.

When studying extreme value statistics we consider the set $S_n = \{x_1, x_2, x_3, \ldots, x_n\}$, where $x_i \stackrel{\text{def}}{=} x(t_i)$, and discuss how the largest value[2]

$$\chi_n = \max[S_n] \tag{5.33}$$

is distributed in the limit of large n. Our heuristic discussion aims at explaining why there are three different classes of asymptotic extreme value distributions and only one asymptote for record statistics. More in depth treatment can be found in Refs. [Gum58, KN00, Gal78, Nev01].

Let $P(x_i)$ denote the common probability density function (PDF) of the x_i's. We furthermore introduce the cumulative distributions

$$F_<(x) = \int_{-\infty}^{x} dx' P(x') \tag{5.34}$$

and $F_>(x) = 1 - F_<(x)$. Since the variables $x_1, x_2, \ldots$ are assumed to be independent,

$$\begin{aligned} H_n(x) &\stackrel{\text{def}}{=} \text{Prob}\{\chi_n < x\} \\ &= \text{Prob}\{x_1 < x, x_2 < x, \ldots, x_n < x\} \\ &= \text{Prob}\{x_1 < x\}\text{Prob}\{x_2 < x\} \cdots \text{Prob}\{x_n < x\} \\ &= [F_<(x)]^n. \end{aligned} \tag{5.35}$$

Since the cumulative distribution satisfies $F_<(x) \leq 1$, Eq. (5.35) implies that for fixed x and in the limit $n \to \infty$ the distribution

[2]The smallest would also do, since a change of sign $x_k \mapsto -x_k$ maps the smallest into the largest value.

$\lim_{n\to\infty} H_n(x)$ approaches either zero or one depending on whether $F_<(x) < 1$ or $F_<(x) = 1$. This indicates that the limit $n \to \infty$ must be treated with care. Mathematical details can be found in the literature, e.g. [Gum58, KN00, Gal78, Nev01]. As the number of elements in the sampling set S_n increases, so will typically their maximum χ_n. This suggests introducing a linear transformation $\chi_n \mapsto (\chi_n - b_n)/a_n$ in a manner similar to what is done to a sum of random variables $S_N = \sum_{i=1}^{N} x_i$, each term having the mean μ and variance σ^2. In that case, $(S_N - \mu N)/(\sqrt{N}\sigma)$ converges to a normal distribution of zero average and unit variance.

Does a limiting distribution $H(x)$ similarly exist for $(\chi_n - b_n)/a_n$ and can we obtain some insight into its properties? The answer is yes, based on the following heuristic argument which goes back to the 1920's [Fré27, FT28]. The idea is based on deriving an equation for the distribution $H_n(a_n x + b_n)$ of the appropriately scaled variable in the limit of large n. The main property used is that the largest value of a union of sets of numbers equals the largest of the maxima of each of the sets.

Consider then N sets each containing n realizations of the stochastic variable x, namely:

$$\{x_1^{(1)}, x_2^{(1)}, \ldots, x_n^{(1)}\},\ \{x_1^{(2)}, x_2^{(2)}, \ldots, x_n^{(2)}\}, \ldots, \{x_1^{(N)}, x_2^{(N)}, \ldots, x_n^{(N)}\}. \tag{5.36}$$

Recalling that $H_n(x)$ is the (cumulative) distribution of the greatest value χ_n, $H_N(a_N x + b_N)$ must then be the distribution function of the greatest value of the combined set

$$S_{Nn} = \{x_1^{(1)}, x_2^{(1)}, \ldots, x_n^{(1)}, x_1^{(2)}, x_2^{(2)}, \ldots, x_n^{(2)}, \ldots, x_1^{(N)}, x_2^{(N)}, \ldots, x_n^{(N)}\}. \tag{5.37}$$

with appropriate choice of a_N and b_N. But since

$$\begin{aligned}\chi_{Nn} &= \max[S_{Nn}]\\ &= \max[\{\max[\{x_1^{(1)}, x_2^{(1)}, \ldots, x_n^{(1)}\}], \ldots,\\ &\quad \times \max[\{x_1^{(N)}, x_2^{(N)}, \ldots, x_n^{(N)}\}]\}]\end{aligned}$$

we arrive at the so-called stability postulate [Fré27, FT28], which reads

$$H_N(a_N x + b_N) = [H_n(x)]^N. \tag{5.38}$$

If we boldly assume that for sufficiently large N and n the functions H_N and H_n can be replaced in Eq. (5.38) by their limit distribution $H = \lim_{m\to} H_m$, we conclude that

$$H(a_N x + b_N) = [H(x)]^N. \tag{5.39}$$

As it turns out, one obtains three different types of extreme value asymptotes depending on the choices of the transformation parameters a_N and b_N. The choice $a_N = 1$ gives the Gumbel type, see Eq. (5.42) below. The two other types correspond to $a_N \neq 1$. These are called Fréchet, see Eq. (5.46) and Weibull, see Eq. (5.48), respectively [Gum58, KN00, Gal78, Nev01]. Before listing all three types, let us briefly discuss how Eq. (5.38) allows one to study the functional form of the limit. We concentrate on the first case where $a_N = 1$ and where the stability condition, Eq. (5.38), reads $[H(x)]^N = H(x + b_N)$. Next we apply Eq. (5.38) to $H(x + b_N)$ and get

$$[H(x)]^{NM} = [H(x + b_N)]^M = H(x + b_N + b_M). \tag{5.40}$$

However, considering that $[H(x]^{MN} = H(x + b_{MN})$ also holds,

$$b_N + b_M = b_{NM}. \tag{5.41}$$

This implies that $b_N = \alpha \log(N)$, where α is a constant. If we choose $\alpha = -1$, Eq. (5.38) is satisfied by $H(x) = \exp(-\exp(-x))$ since then

$$H^N(x) = \exp(-N \exp(-x))$$

and

$$\begin{aligned} H(x - \log(N)) &= \exp(-\exp(-x + \log(N))) \\ &= \exp(-N \exp(-x)) = H^N(x). \end{aligned}$$

The three different types of asymptotic distributions are given by:

1. Gumbel type

$$H(x) = \exp[-\exp(-x)] \tag{5.42}$$

where $-\infty < x < \infty$. The Gumbel limit applies when the probability density function $P(x)$ for the stochastic variables x_i vanishes sufficiently fast for large x, (see [Gal78]). More precisely, when for some finite value a, the integral

$$\int_a^{\omega(F)} (1 - F(x))dx \tag{5.43}$$

is finite. The upper limit of the integral is given by $\omega(F) = \sup\{x : F(x) < 1\}$. Let $\alpha(F) \stackrel{\text{def}}{=} \inf\{x : F(x) > 0\}$ and for $\alpha(F) < t < \omega(F)$ let

$$R(t) = \frac{\int_t^{\omega(F)} (1 - F(x))dx}{1 - F(t)}. \tag{5.44}$$

The form in Eq. (5.42) applies when for all $x \in \mathbb{R}$

$$\lim_{t \to \omega(F)} \frac{1 - F(t + xR(t))}{1 - F(t)} = e^{-x}. \tag{5.45}$$

2. Fréchet type

$$H(x) = \begin{cases} \exp(-x^{-\gamma}) & \text{if} \quad x > 0 \\ 0 & \text{if} \quad x \leq 0. \end{cases} \tag{5.46}$$

The exponent is determined from the limit

$$\lim_{t \to \infty} \frac{1 - F(tx)}{1 - F(t)} = x^{-\gamma}. \tag{5.47}$$

3. Weibull type

$$H(x) = \begin{cases} 1 & \text{if} \quad x > 0 \\ \exp[-(-x)^{\gamma}] & \text{if} \quad x \le 0. \end{cases} \tag{5.48}$$

The exponent is in this case determined by the limit

$$\lim_{t\to\infty} \frac{1 - F(\omega(F) - \frac{1}{tx})}{1 - F(\omega(F) - 1/t)} = x^{-\gamma}. \tag{5.49}$$

Here the 'end point' of the distribution $F(x)$, given by $\omega(F) = \sup\{x : F(x) < 1\}$, is assumed to be finite.

The normalization coefficients a_n and b_n are not unique and possible choices are discussed in [Gal78] and [Gum58]. Furthermore, the range of attraction of the three different types of asymptotes is still a matter of research. Rather than entering into these matters, we will just consider some typical examples.

Exponential probability density. Assume the probability density function $f(x)$ for the individual stochastic variables x_i to be given by

$$f(x) = \begin{cases} e^{-x} & \text{for} \quad x \le 0 \\ 0 & \text{for} \quad x < 0. \end{cases} \tag{5.50}$$

The corresponding cumulative distribution function is $F(x) = \int_0^x e^{-x'} dx' = 1 - e^{-x}$. The condition described in Eq. (5.43) is fulfilled since $\omega(F) = \infty$ and

$$\int_0^\infty [1 - F(x)]dx = \int_0^\infty e^{-x} dx = 1 < \infty.$$

The quantity $R(t)$ in Eq. (5.44) becomes

$$R(t) = \frac{\int_t^\infty e^{-x} dx}{e^{-t}} = 1.$$

Hence the exponential PDF falls within the Gumbel class.

Power-law probability density. Assume that the probability density function, or PDF, is given by

$$f(x) = \begin{cases} (1-a)x^{-a} & \text{for} \quad x \geq 1 \\ 0 & \text{otherwise.} \end{cases} \tag{5.51}$$

We assume $a > 1$ to ensure normalizability, i.e. the convergence of $\int_1^\infty f(x)dx$. The cumulative distribution

$$F(x) = \begin{cases} 0 & \text{for} \quad x < 1 \\ \int_1^x (1-a)t^{-a}dt = 1 - x^{1-a} & \text{for} \quad x \geq 1, \end{cases} \tag{5.52}$$

is characterized by $\omega(F) = \infty$ and fulfils

$$\frac{1 - F(tx)}{1 - F(t)} = \frac{1 - (1 - (tx)^{1-a}}{1 - (1 - t^{1-a})} = x^{-(a-1)}, \tag{5.53}$$

i.e. the assumed power-law distribution corresponds to the Fréchet case with $\gamma = a - 1$.

Bounded probability density. Next, as an example of a PDF supported on a bounded interval consider

$$f(x) = \begin{cases} 1 & \text{for} \quad x \in [0,1] \\ 0 & \text{otherwise.} \end{cases} \tag{5.54}$$

The cumulative distribution is given by

$$F(x) = \begin{cases} 0 & \text{for} \quad x < 0 \\ x & \text{for} \quad x \in [0,1] \\ 1 & \text{for} \quad x \geq 1. \end{cases} \tag{5.55}$$

We have $\omega(F) = 1$, and to apply Eq. (5.49) we need

$$F^*(x) = F(1 - 1/x) = \begin{cases} 1 & \text{for} \quad x \leq 0 \\ 0 & \text{for} \quad 0 < x < 1 \\ 1 - 1/x & \text{for} \quad x \geq 1 \end{cases} \tag{5.56}$$

and hence for large t

$$\frac{1 - F^*(tx)}{1 - F^*(t)} = \frac{1 - (1 - \frac{1}{tx})}{1 - (1 - \frac{1}{t})} = \frac{1}{x}. \tag{5.57}$$

From which we conclude that we are dealing with a Weibull or type 3 case with $\gamma = 1$ in Eq. (5.49).

5.4.1. Relation between Record and Extreme Value Statistics

From the arguments above it is clear that, even though both concerned with the largest value of some set of data, the statistics of records and that of extremes are quite different. The first deals with *when* a new record occurs, and implies a time ordering of the data. The second focusses on the size distribution of the largest value in a set. Properties of record statistics can be obtained from those of extreme value statistics through the following observation: Consider a sequence

$$\nu_0, \nu_1, \nu_2, \nu_3, \ldots, \nu_N, \nu_{N+1}, \ldots$$

of independent and identically distributed stochastic variables. The probability P_N that ν_N is a record can be expressed in the following way

$$P_N = \int_{-\infty}^{\infty} \text{Prob}\{\max[\nu_0, \nu_1, \nu_2, \ldots, \nu_{N-1}] < x\} P_\nu(x)dx, \tag{5.58}$$

where $P_\nu(x)$ is the probability density shared by all the ν_k's. As we did in Eq. (5.35), we notice that

$$\text{Prob}\{\max[\nu_0, \nu_1, \nu_2, \ldots, \nu_{N-1}] < x\} = [F_\nu(x)]^N, \tag{5.59}$$

where $F_\nu(x)$ of is the cumulative distribution for one of the ν variables. Furthermore, since $P_\nu(x) = dF_\nu(x)/dx$ one obtains

$$P_N = \int_{-\infty}^{\infty} [F_\nu(x)]^N \frac{dF_\nu(x)}{dx} dx = \int_{-\infty}^{\infty} F_\nu^N dF_\nu = \frac{1}{N}, \tag{5.60}$$

i.e. the same result as in Eq. (5.1). The derivation makes it explicit that although extreme value statistics depends on the nature of the probability density function of the underlying stochastic variables, the probability of a record does not. For the same reason it may be illuminating to repeat the calculation of Sec. (5.2) starting from extreme value considerations. Assume then that records occur at time steps m_i with $i = 1, 2, \ldots, n-1$ in addition to the record at time step zero. We can express $P_n(l, [0, m_1, m_2, \ldots m_{n-1}])$ (see Eq. (5.6)) as

$$P_n(l, [0, m_1, m_2, \ldots, m_{n-1}]) = \int_{-\infty}^{\infty} dx_{n-1} P_\nu(x_{n-1})$$
$$\int_{-\infty}^{x_{n-1}} dx_{x_{n-2}} P_\nu(x_{x_{n-2}}) \int_{-\infty}^{x_{n-2}} dx_{n-3} P_\nu(x_{n-3}) \cdots$$
$$\int_{-\infty}^{x_{m_2}} dx_{m_1} P_\nu(x_{m_1}) \int_{-\infty}^{x_{m_1}} dx_0 P_\nu(x_0)$$
$$\text{Prob}\{x_1 < x_0, x_2 < x_0, \ldots, x_{m_1-1} < x_0\}$$
$$\text{Prob}\{x_{m_1+1} < x_{m_1}, x_2 < x_{m_1}, \ldots, x_{m_2-1} < x_{m_1}\}$$
$$\ldots$$
$$\text{Prob}\{x_{m_{n-2}+1} < x_{m_{n-2}}, \ldots, x_{m_{n-1}-1} < x_{m_{n-2}}\}$$
$$\text{Prob}\{x_{m_{n-1}+1} < x_{m_{n-1}}, \ldots, x_l < x_{m_{n-1}}\}. \tag{5.61}$$

We now introduce the distribution function, simplify the notation for the integration variables and obtain

$$P_n(l, [0, m_1, m_2, \ldots, m_{n-1}]) = \int_{-\infty}^{\infty} dx_{n-1} P_\nu(x_{n-1})$$
$$\int_{-\infty}^{x_{n-1}} dx_{x_{n-2}} P_n u(x_{x_{n-2}}) \int_{-\infty}^{x_{n-2}} dx_{n-3} P_\nu(x_{n-3}) \cdots$$
$$\int_{-\infty}^{x_{m_2}} dx_{m_1} P_\nu(x_{m_1}) \int_{-\infty}^{x_{m_1}} dx_0 P_\nu(x_0)$$
$$F_\nu(x_{m_{n-1}})^{l-m_{n-1}} F_\nu(x_{m_{n-2}})^{m_{n-1}-m_{m-2}-1} F_\nu(x_{m_{n-3}})^{m_{n-2}-m_{n-3}-1}$$
$$\cdots F_\nu(x_{m_1})^{m_2-m_1-1} F_\nu(x_0)^{m_1-1}. \tag{5.62}$$

Making the substitution $z_i = F_\nu(x_i)$, which we can do since $P_\nu(x_i) = dF_\nu/dx_i$, the above expression reduces to the integral

$$P_n(l, [0, m_1, m_2, \ldots, m_{n-1}])$$
$$= \int_0^1 dz_{n-1} \int_0^{z_{n-1}} dz_{n-2} \cdots \int_0^{z_2} dz_1 \int_0^{z_1} dz_0 \, z_{n-1}^{l-m_{n-1}}$$
$$\times z_{n-2}^{m_{n-1}-m_{n-2}-1} z_{n-3}^{m_{n-2}-m_{n-3}-1} \cdots z_{m_1}^{m_2-m_1-1} z_0^{m_1-1}. \tag{5.63}$$

The derivation shows why $P_n(l, [0, m_1, m_2, \ldots, m_{n-1}])$ is independent of the distribution of the signal from which the records are extracted. The integrals can easily be performed by starting with the integral over z_0, followed by integration over z_1, and so forth. The result is

$$P_n(l, [0, m_1, m_2, \ldots, m_{n-1}]) = \frac{1}{l m_{n-1} m_{n-2} \cdots m_1 m_0}, \tag{5.64}$$

which is in agreement with Eq. (5.5).

5.5. SUMMARY

In summary, the statistics of extreme values is on the one hand more general and on the other more specific than that of record values. Firstly, it is not concerned with the order in which the extremes occur and, for this reason, record statistics can be derived from it. Secondly, it yields the probability distribution of the values of the extremes, whereas record statistics is indifferent to it. Being concerned with the actual values of the extremes, extreme value statistics is sensitive to the distribution of the underlying stochastic variables. This leads to three different universality classes for the distribution of extrema. In contrast, record statistics is essentially described by the log-Poisson process irrespective of the distribution.

6

Complexity and Hierarchies

6.1. INTRODUCTION

Natural and man-made systems range from the very simple to the very complex. A spinning top is easily described in terms of its spinning, precession and nutation frequencies, while the motion of a turbulent fluid has a wide spectrum of length and time scales associated to its eddies. Molecular motion in a gas enclosed in a container is usually highly chaotic. Yet, this very feature allows a statistical treatment where only few variables appear, e.g. prominently, the average distance travelled by a molecule between collisions and the molecule's mean free path. Whenever dominant characteristic length, energy and time scales can be identified, the corresponding phenomena can be studied in relative isolation, e.g. biochemical reactions inside an organism can be understood without worrying about nuclear reactions inside a star. If many different scales are intermixed, structural and dynamical features of the system as a whole do matter. This is where complexity and hierarchies enter the scene.

Two different but complementary issues arise in complex system dynamics. The first is explaining the ability of certain complex molecules, or systems of molecules, to organize themselves into ordered structures of great stability in the face of thermal noise and of a large number of nearly equivalent available configurations, i.e. how can a protein quickly fold into its native structure?

The prediction of the structure and function of chemical compounds, based on accurate models of the interactions of its component atoms, is in the same general category. The second issue regards the opposite phenomenon, namely the persistent effects on the dynamics of the initial configuration. The phenomenon is observed in glassy systems undergoing a prolonged non-stationary *ageing* process where multiple metastable configurations are visited in a certain order.

The connection between complexity and hierarchies was emphasized long ago by Herbert A. Simon [Sim62] in his seminal essay "The Architecture of Complexity". Simon defines a hierarchy as a system composed of subsystems, each again composed of subsystems, as in a set of Russian matryoshka dolls. Many physical, biological and social systems conform to this paradigm, e.g. individuals are the basic 'units' of social structures comprising couples, families, clans, societies and civilizations. From a biological point of view, an individual is at the top level of a hierarchy comprising organs, tissues, cells, and organelles. The subunits of a hierarchy can be defined in terms of characteristic length or time scales, or in terms of their interactions with other subunits. In any case, a process initiated within a subunit must, in the short run, remain confined to that subunit, and only in the long run be able to affect the whole structure. The dynamics is then nearly *decomposable*.

Complex systems emerge in a robust and relative expedite fashion through an iterated process of assembly and adaptation of pre-existing elements. The relevance of iteration in biological evolution is hinted to by a degree of similarity between ontogeny, the sequence of events involved in the development of an individual organism, and phylogeny, the origin and evolution of a species.

As hierarchies generally embody a structure which implies a partial ordering of the components of the hierarchy, trees, of the sort shown in Fig. (6.1) above, provide a natural description. Trees are graphs whose elements, or nodes, are connected by a unique sequence of edges, or *path*. *Rooted trees* are trees endowed with a *partial ordering*: For any two nodes a and b, $a \geq b$ if and only if the unique path connecting the root node to node b passes

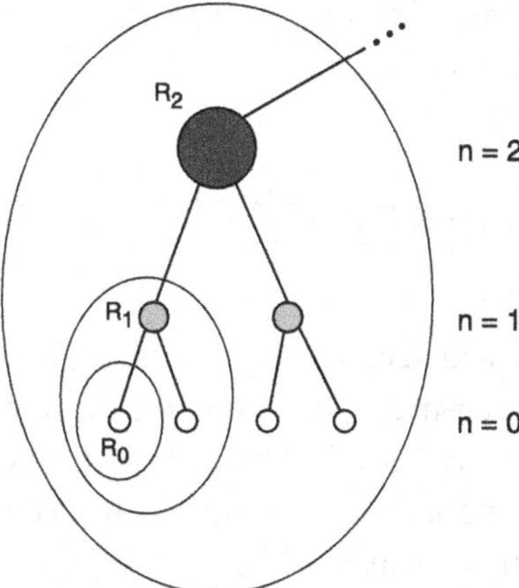

Figure 6.1. A tree model with three levels is shown. The size of the spheres corresponds to the degeneracy of the nodes. The nested contours indicate the ergodic components successively reaching local equilibrium as the system relaxes from its initial state in the leftmost node.

through node *a*. Phylogenetic trees and organizational charts are both examples of rooted trees illustrating a set of hierarchical relationships: The position of biological species in a time ordered evolutionary process in the first case, and the command chain of a company in the second.

In this chapter, the words tree, rooted tree and hierarchy are used interchangeably. Two different but related types of hierarchical dynamics are discussed: In the first, the system variables are linked by a set of kinetic constraints, i.e. a variable placed at level k in a tree is unable to change its state, unless a certain subset of variables at level $k-1$ has reached a specific configuration. This is similar to an organizational chart, except that the information controlling the dynamics goes from the bottom to the top of the hierarchy, and not from the top to the bottom as in a chain of command. In the second type, a tree structure arises when coarse-graining the configuration space, also called the *energy landscape*, of a thermally relaxing complex system. The nodes of the tree represent in this case sets of metastable configurations which are able to support a state of near thermal equilibrium, e.g. for a configuration space representing different arrangements of interacting atoms a node would represent a metastable chemical compound.

6.2. KINETICALLY CONSTRAINED HIERARCHICAL DYNAMICS

Kinetic constraints are dynamical rules forbidding certain transitions, without necessarily affecting equilibrium properties, e.g. Conway's "Game of Life" is a dynamical system controlled by kinetic constraints. Kinetic constrained models describe the behaviour of dense glass forming liquids [KA93], e.g. the divergence of the diffusion coefficients for increasing density. Here we are concerned with the kinetically constrained hierarchical models described in Ref. [PSAA83]. The toy model discussed below allows a semi-analytic treatment of hierarchical dynamics with kinetic constraints. Secondly, it provides the opportunity to introduce the concept of local thermal equilibration, which is used later in this chapter.

With reference to Fig. (6.1), we assume that a dichotomic variable, with values 0 or 1, resides on each of the nodes of a binary tree having N levels, the latter numbered from 0 to $N-1$. At level 0, the 2^N variables $x_0^0(t), x_1^0(t) \dots x_{2N-1}^0$ flicker randomly and independently at a rate μ between their two states. First level variables $x_0^1(t), \dots x_{2N-1-1}^1$ also flip at a rate μ, but only if their two 'daughter' variables both are in state 1, i.e. $x_l^1(t)$ can only change its state when $x_{2l}^0(t) = 1$ and $x_{2l+1}^0(t) = 1$.

The relation between the dynamics at levels 0 and 1, coupled with the hierarchical organization of the constraints, suffices to derive the general properties in a tree with an arbitrary number of levels.

Since the variables are identically distributed and independent at each level, it is sufficient to analyse the dynamical evolution of one of them. At level zero we shall hence consider x_0^0. Let $T_{ii}^0(t)$, where $i = 0$ or 1, denote the probability that $x_0^0(t)$ be in the same state i at times 0 and t, i.e. the probability that an even number of flips has occurred in the intervening time. The number of flips is

Poisson distributed, and the required probability is hence given by the sum

$$T_{ii}^{0}(t) = e^{-\mu t} \sum_{l=0}^{\infty} \frac{(\mu t)^{2l}}{(2l)!} = e^{-\mu t} \cosh(\mu t) = \frac{1}{2}(1 + e^{-2\mu t}). \quad (6.1)$$

Similarly, the variable will be in different states at times 0 and t if an odd number of flips has occurred in $(0, t)$. The elements of the propagator $\mathbf{T}$ can hence be written concisely as

$$T_{ik}^{0}(t) = \frac{1}{2}(1 - e^{-2\mu t}) + \delta_{ik} e^{-2\mu t}, \quad i, k \in \{1, 2\}. \quad (6.2)$$

Referring to Eq. (3.85), and considering that the equilibrium probability $P_{\mathrm{eq}}^{0}(k)$ equals $1/2$, the autocorrelation function for x_0 can be written as

$$\mathrm{Corr}_{x_0^0}(t) = \frac{1}{2} \sum_{i=0}^{1} \sum_{k=0}^{1} \left(i - \frac{1}{2}\right)\left(k - \frac{1}{2}\right) T_{ik}^{0}(t). \quad (6.3)$$

Exercise 1: Verify Eq. (6.3) and show that the equation can be rewritten as

$$\mathrm{Corr}_{x_0^0}(t) = \frac{1}{4}\left(T_{11}^{0}(t) - T_{10}^{0}(t)\right) = \frac{1}{4} e^{-2\mu t}. \quad (6.4)$$

The expression is obtained by weighing the initial states 0 and 1 equally, i.e. according to their equilibrium probabilites. Show that the result also applies when the initial state is 0 (or 1) with probability 1. Note that the pre-factor $\frac{1}{4}$ equals the equilibrium variance of x_0.

Exercise 2: Show that the autocorrelation of the vector-valued process $\mathbf{x}^0(t) = [x_0^0(t), x_1^0(t), \ldots x_{2^N-1}^0(t)]$ is given by

$$\mathrm{Corr}_{\mathbf{x}^0}(t) = 2^{N-2} e^{-2\mu t}. \quad (6.5)$$

To discuss the dynamical relaxation of the whole hierarchy, we start considering $x_0^1(t)$, the first level variable linked to $x_0^0(t)$ and $x_1^0(t)$. Assuming that the latter variables are initially in states l and m, the propagator of x_0^1, $\mathbf{T}^1$, obeys the master equation

$$\frac{dT^1_{ik|lm}}{dt}(t) = \mu T^0_{1,l}(t)T^0_{1,m}(t)\left(1 - 2T^1_{ik|lm}(t)\right). \tag{6.6}$$

To enforce the kinetic constraint, the transition rate μ has been rescaled, on the r.h.s. of the equation, by the probability that both 'daughter' variables be in state 1. Secondly, the normalization condition $T^1_{1k|lm}(t) + T^1_{0k|lm}(t) = 1$ has been used to obtain a closed form equation. Equation (6.6) is a first order differential equation, with solution

$$T^1_{ik|lm}(t) = \frac{1}{2}\left(1 - e^{-2\mu\int_0^t T^0_{1,l}(t')T^0_{1,m}(t')dt'}\right) + \delta_{ik}e^{-2\mu\int_0^t T^0_{1,l}(t')T^0_{1,m}(t')dt'}. \tag{6.7}$$

Note the structural similarity to Eq. (6.2). Next, we consider the initial condition where all variables are initially in state 0, i.e. a configuration fully constrained at all levels higher than zero. The gradual thawing of the system is expressed by the configurational autocorrelation function. To simplify the rather cumbersome integrals present in Eq. (6.7), we note that for times larger than $\tau_0 = \mu^{-1}$, equilibrium is reached at the lower level, and the different elements of $\mathbf{T}^0$ can all be approximated by 1/2. Equivalently, the transition rate in equation (6.6) can be replaced by $\mu/4$, i.e. moving one level up in the hierarchy, the transition rate is reduced by a factor of 1/4. On the same time scale, variables at two different levels decorrelate. By the same token, the transition rate for level k variables is $\mu 2^{-2k}$ on time scales larger than $\tau_{k-1} = 2^{2k-2}/\mu$. Letting $\mathbf{x}$ denote the full set of $2^{N+1} - 1$ variables, ordered layer by layer, the autocorrelation of $\mathbf{x}$ is approximately equal to a sum of the contributions from each level:

$$\mathrm{Corr}_{\mathbf{x}}(t) = \sum_{k=0}^{N} 2^{N-2-k}e^{-\mu 2^{-2k+1}t}. \tag{6.8}$$

Normalizing to one at $t = 0$, and taking the limit $N \to \infty$, the autocorrelation function takes the form

$$C_{\mathbf{x}}(t) = \frac{1}{2}\sum_{k=0}^{\infty} 2^{-k} e^{-\mu t 2^{-2k+1}}. \tag{6.9}$$

Exercise 3: Consider a rescaling of the time variable of the form

$$t \to r_l t, \quad \text{where } r_l = 2^{-2l}, \quad \text{and where } l = 1, 2, 3, \ldots. \tag{6.10}$$

Show that

$$C_{\mathbf{x}}(r_l t) = (r_l)^{-1/2} C_{\mathbf{x}}(t). \tag{6.11}$$

Rescaling time thus amounts to a rescaling of the autocorrelation function. If the procedure worked for all real values of r, $C_{\mathbf{x}}(t)$ would, asymptotically for large t, approach the power-law $t^{-1/2}$, for $t \gg 0$. Due to the discrete structure of our tree, only a discrete set of rescalings works in our case. To what extent does $C_{\mathbf{x}}(t)$ then resemble a power-law?

Exercise 4: Discuss how the arguments presented in this section must be modified, if the branching ratio of the tree, i.e. the number of 'daughter' nodes is 3 rather than 2.

The model just discussed may appear somewhat artificial, since the hierarchy of variables it introduces could be hard to identify in an actual physical system. The hierarchy of relaxation time scales which the model produces is nevertheless present in many complex dynamical systems.

The configuration space of kinetically constrained models is akin to a maze, where the difficulty of moving around does not stem from the necessity of climbing high energy barriers, but rather reflects the difficulty of finding the correct path in a vast configuration space. This is akin to finding the proverbial 'needle in a haystack', and the impediment to a quick solution to the problem is often called an *entropic barrier*. In the following example, the typical time required to exit a certain region of a maze increases exponentially with the linear size of the region, in the same way as the equilibration times of the hierarchical model just discussed.

Example 1: A tourist in a maze is faced with a number of binary choices, i.e. he must decide whether to turn or not when passing a gate. The number of gates in an area of linear size L is proportional to L. The possible number of paths is then $\exp(cL)$, where c is a constant. Of all these paths, only one leads to the exit from the area. Assuming that the tourist keeps track of his steps and avoids repeating the same mistakes, the time required to reach the exit typically involves exploring half of the available possibilities. Consequently, the time scale for exiting the area is $\tau_L = \exp(cL)/2$.

6.3. DYNAMICAL HIERARCHIES IN THERMAL RELAXATION

In a thermally activated relaxation process, or *thermalization*, the configuration energy only enters divided by the temperature of the thermal bath. Hence, energy 'bumps' of size e will not be visible on the energy hyper-surface of the system for $T \gg e$. For the same reason, unseen features, e.g. new local energy minima, may appear as the temperature is lowered. A hierarchy of nested local minima therefore naturally emerges upon cooling, with the temperature acting as an adjustable low energy cut-off. At a fixed low temperature, the relaxation goes through a sequence of near equilibrium states, which are supported in nested subsets of *ergodic components*, $\mathcal{R}_1 \subset \mathcal{R}_2 \ldots \subset \mathcal{R}_n, \ldots$. This paradigm also applies to relaxation mechanisms where the barriers confining the ergodic components are not energy but entropy barriers, as in the maze example discussed above.

In the next chapter, we discuss how to coarse-grain the thermalization dynamics in the energy landscape of a complex system. At this stage we only investigate the scaling properties of relaxation which can be derived based on simple heuristic assumptions. The general assumption concerns the hierarchical nature of the process, i.e. the fact that it proceeds in stages corresponding to progressively higher levels of the hierarchy being visited with increasing probability. More specifically we assume that all microscopic configurations present in the n'th ergodic component have equal probability:

The latter is zero if the component has not yet been reached by the relaxation process starting at state k. At later times, for any $x \in \mathcal{R}_n$ it is given by $P_k(x,t) = 1/V_n$, where V_n is the 'volume' of the n'th component, i.e. the number of microscopic configuration it contains. Clearly, this approximation does away with all internal relaxation processes within components. Finally, we assume that equilibrium at stage n is reached on a time scale t_n. Both V_n and t_n are growing functions of n, and $n = n(t)$ can hence be considered to be a function of time. Choosing different n dependencies for V_n and t_n allows us to explore a wide range of possible relaxation behaviours.

To construct a (very coarse) approximation for the propagator, we let $n^*(i,k)$ be the index of the smallest ergodic component containing both state i and state k in a sequence of components starting with state k. The probability $T_{ik}(t)$ of the system being in state i at time t, given initial state k is then given by

$$T_{ik}(t) = 0, \quad \text{for } t < t_n^*(i,k) \tag{6.12}$$

$$T_{ik}(t) = V_{n^*(i,k)}, \quad \text{for } t > t_n^*(i,k). \tag{6.13}$$

Consider now the case where $t_n = \tau \exp(an)$, $V_n = V_0 \exp(bn)$, and where a, b, τ, and V_0 are real and positive constants. The probability $T_{kk}(t)$ of being at the same state at times 0 and t is then

$$T_{kk}(t) = \left(\frac{t}{\tau}\right)^{-b/a}, \quad t > \tau. \tag{6.14}$$

The approximation scheme has no prediction for what happens for $t < \tau$, since τ is the relaxation time in the initial component, and all internal relaxation processes are neglected. The probability of being in state k is seen to decay algebraically in time. The exponents depend on the geometrical properties of the configuration space through the constant a and b.

Exercise 4: Verify Eq. (6.14). Note that while a power-law is invariant under a continuous scale transformation, the arguments leading to Eq. (6.14) only apply to a discrete set of exponentially separated times $t_1, t_2 \ldots t_n$.

Exercise 5: Repeat the steps leading to Eq. (6.14) for $t_n = \tau \ln(an)$ and $V_n = V_0 \exp(bn)$.

In conclusion, varying some simply stated scaling assumptions on geometrical properties of configuration space, it easy to map out the kind of relaxation behaviour which can qualitatively be expected in a hierarchical organized configuration space. Whether the assumptions are at all correct for a given model or situation is, of course, a different matter altogether.

7

Energy Landscapes

7.1. INTRODUCTION

Commonly used in physics and chemistry, the term 'energy landscape' is probably derived from the analogous 'fitness landscape' earlier introduced in evolutionary biology by Sewall Wright [Wri86]. It evokes the picture of a mountain range, positions and height respectively corresponding to microscopic configurations of the system considered and to their energies. More precisely, what defines an energy landscape is the set of configurations of a physical system equipped with a real valued energy function and with a rule specifying which states are neighbours. The stochastic dynamics we consider below involves thermally activated transitions between neighbouring states.

Thermal relaxation dynamics in energy landscapes with a large number of metastable configurations is important not only from an application point of view, but also because it offers theoretical insights which can be transferred to different problems, e.g. to population dynamics, where the source of noise is not thermal, but rather linked to random mutations occurring during the replication of organisms.

Discrete energy landscapes can be viewed as graphs whose nodes represent configurations and whose edges represent connections between neighbouring nodes. The distance between two configurations is defined as the smallest number of edges needed to connect the corresponding nodes. Continuous energy landscapes are endowed with the usual Euclidean metrics in $\mathcal{R}^n$. Dynamical transitions from a point $\mathbf{x}$ are in this case associated to an *a priori* probability, which falls off with the distance between the pair of points involved. In some cases, the probability is taken to vanish outside a ball centred at $\mathbf{x}$.

Example 1: The ferromagnetic Ising spin model mentioned in Example (3.3) has N spin placed on a regular grid. The model has the hypercube as its configuration space, with adjacent corners of the hypercube representing neighbouring configurations. The spin glass model with the energy function defined in Eq. (9.1) has the same configuration space but a different landscape, as the energy function is different: In the first case the interaction between neighbouring spins is ferromagnetic, and in the second it varies in sign and intensity in a manner which is random but fixed, for a given sample.

A set of N particles moving in three spatial dimensions has $\mathcal{R}^{3N}$ as configuration space, and the usual Euclidean metrics defines the topology. Together with an energy function, e.g. a sum of pairwise interactions between particles, this defines a continuous energy landscapes. Importantly, the momenta of the particles are not included in the description. This is tantamount to a *fully damped* description, a description where inertial terms have no relevance. Transition rates between any two microscopic states in the landscape only depend on their energy difference, as required by detailed balance, see Eq. (3.67), and knowledge of the energy function fully specifies the dynamical evolution of the system.

Many energy landscapes are closely associated to models of compounds and materials, and knowledge of their structure [Wal03] plays an increasing role in solid state chemistry and physics, e.g. the thermal properties of large molecules, solids and magnetic systems are often modelled in terms of energy landscapes. From our present perspective, central features of glassy dynamics are closely related to landscape geometry, i.e. hierarchical energy landscapes provide a useful link to theory.

The first part of this chapter focusses on models and on the mathematical techniques required to solve them. The second on how to use numerical techniques to extract the geometrical properties of landscapes which are, in a large class of systems, the basis of complex dynamics.

7.2. MODELLING ENERGY LANDSCAPES

The landscape concept hides a considerable complexity since, for every configuration of the component particles, one has to calculate the potential energy, usually, but not exclusively, modelled in terms of pairwise interactions between particles. Once this is done, reduced descriptions with predictive power must be constructed. In this section, we are mainly interested in how the coarse-graining process in principle can be carried out, and less interested in detailed applications. Hence our focus on simple toy models which in spite of their simplicity capture the main characteristics of complex thermal behaviour.

7.2.1. Coarse-graining and the Kramers' Approximation

Coarse-graining is commonly and often tacitly used to remove details deemed irrelevant on the length and/or time scales of interest and produce an effective description with better predictive power. The first coarse-graining step is the elimination of all inertial terms, a step already implicit in the landscape description. The second is the Kramers' approximation [Kra40], originally developed for chemical reaction kinetics, but generally applicable to thermalization processes.

> *Example 2:* The relative motion of the molecules in a 'rigid' body is safely swept under the carpet of the body's alleged rigidity. If we allow the body to flex, the motion can be described using waves in an elastic continuum, still a huge simplification. Any macroscopic description relies on coarse-graining.

Example 3: Consider the motion of an interstitial point defect in a crystal. The defect is described by quantum mechanics on very short time scales. On intermediate time scales, it is trapped near a definite interstitial position corresponding to a minimum of the potential energy. On longer scales, it can jump from one position to a neighbouring position by surmounting an energy barrier of magnitude b. The symmetry of the crystal lattice implies that all barriers separating neighbours have the same magnitude. Hence, even though the system may possess a very large number of potential energy minima, defect motion is simply diffusive. The process is time homogeneous, and hardly qualifies as 'complex'. The only aspect of the microscopic description surviving at the macroscopic level is the energy barrier b separating neighbouring sites. The barrier enters the expression for the diffusion constant $D \propto e^{-b/T}$, where T is the temperature and where the Boltzmann constant is, as usual, set to one.

In an energy landscape of any dimensionality, the configurations are denoted by $\mathbf{x}$, and the energy function by $E(\mathbf{x})$. When $E(\mathbf{x})$ possesses precisely two minima, the catchment basins or *valleys*, $\mathcal{V}_i$, $i = 1$ or 2, can be defined as the sets of points from which trajectories flow toward either minimum, in the limit $T \to 0$, i.e. the limit where all 'uphill' dynamical transitions are forbidden. The watershed line between the two valleys usually passes through one or more saddle points. The saddle point of lowest energy dominates the inter-valley transitions at low T, and is therefore the only one considered in the following. Without loss of generality, its position is taken to be at $\mathbf{x} = 0$.

The Kramers' method relies on a coarse-graining technique: A state of 'local' thermal equilibrium establishes itself in each valley on a time scale much shorter than the time typically needed to cross the saddle point. To explore this idea, we define *indicator* functions I_i by $I_i(\mathbf{x}) = 1$ if $\mathbf{x} \in \mathcal{V}_i$ and $I_i(\mathbf{x}) = 0$ otherwise. Secondly, we let $P_{\mathrm{eq},i}$ denote the Boltzmann equilibrium distribution in valley i. Dynamically, local equilibrium is reached if reflecting boundaries are imposed on the perimeter of the valleys. In practice, we just need to use the properly normalized restriction of the Boltzmann equilibrium of either valley:

$$P_{\text{eq},i}(\mathbf{x}) = \frac{e^{-\frac{E(\mathbf{x})}{T}}}{Z_i}, \tag{7.1}$$

where Z_i is the partition function of valley i. Thirdly, we let $p_i(t)$ denote the time dependent probability of finding the system within valley i. Neglecting all the 'fast' relaxation processes implies that the probability flow across the saddle point 'immediately' redistributes itself according to local thermal equilibrium. Hence,

$$P(\mathbf{x}, t) = \sum_{i=1}^{2} I_i(\mathbf{x}) P_{\text{eq},i}(\mathbf{x}) p_i(t). \tag{7.2}$$

The change of $p_i(t)$ is controlled by the transition rates between the valleys,

$$k_{12} = vP_{\text{eq},2}(\mathbf{0}) \quad \text{and} \quad k_{21} = vP_{\text{eq},1}(\mathbf{0}), \tag{7.3}$$

where the constant v provides the correct physical dimension. Accordingly, the coarse-grained master equation is

$$\frac{dp_1(t)}{dt} = -k_{21}p_1(t) + k_{12}p_2(t) \tag{7.4}$$

$$\frac{dp_2(t)}{dt} = k_{21}p_1(t) - k_{12}p_2(t), \tag{7.5}$$

with rate coefficients given in Eq. (7.3). In the special case where the two valleys are identical, the energy of both minima can be taken to be zero. Denoting the energy of the saddle by b, the 'barrier' energy, we find

$$k_{12} = k_{21} = \frac{\exp\left(\frac{-b}{T}\right)}{Z}. \tag{7.6}$$

Exercise 1: Equation (7.6) can be rewritten as

$$k_{12} = k_{21} = \exp\left(\frac{-\Delta F}{T}\right), \tag{7.7}$$

where ΔF is the *free energy barrier* between the two valleys. Express ΔF as a function of Z and T and verify that the result conforms to the standard definition of free energy in equilibrium statistical mechanics.

The eigenvalues of the master equation (7.5) are the equilibrium eigenvalue $\lambda_1 = 0$ and the relaxation eigenvalue $\lambda_2 = -\frac{2\exp(\frac{-b}{T})}{Z}$. The relaxation time

$$\tau = -\frac{1}{\lambda_2} = \frac{Z}{2}\exp\left(\frac{-b}{T}\right) \tag{7.8}$$

is the equilibration time scale of the bistable system. If valleys lack internal energy barriers, their equilibration times have no strong T dependence. At low T, the internal equilibration will then be completed on a time scale much smaller than τ, confirming the internal consistency of the Kramers' approximation.

Exercise 2: Verify the results given above regarding the thermalization of a bistable system.

Exercise 3: In the general case, find the two eigenvalues of the coarse-grained master equation, Eq. (7.5). Comment on the dependence of the relaxation eigenvalue on the 'form' of each valley, as expressed by the respective partition function. Find the equilibrium probability of finding the system in either valley.

Exercise 4: In a one-dimensional toy model valley 1 is the interval $[-1, 0)$ and valley two is the interval $(0, 1]$. The energy is $E(x) = 0$ in the first valley and $E(x) = m$ in the second, i.e. the valleys are 'flat'. At the boundary point the energy is $E(0) = 2m$. Find the time dependence of $p_2(t)$, given that the system is initially in either valley 1 or valley 2.

Exercise 5: Argue that the motion of the interstitial defect described in Example 3 is diffusive and that the diffusion constant has the form $D = D_0 \exp(\frac{b}{k_B T})$, where T is the temperature and k_B is the Boltzmann constant. How would you phenomenologically describe the probability density that a defect initially located at an interstitial site has not left the site at time t?

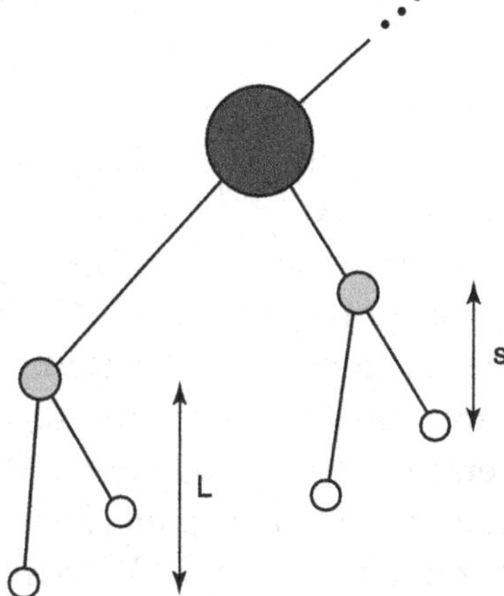

Figure 7.1. A tree model with dynamically inequivalent minima.

7.2.2. Applying the Kramers' Approximation to a Hierarchical Landscape Model

Coarse-graining often results in a master equation with states connected in a tree structure. Here, we consider thermal relaxation on such structures, exemplified by the hierarchical tree model illustrated in Fig. (7.1) and first introduced in Ref. [SH91]. In spite of its outmost simplicity, this model is already useful to describe thermal relaxation in complex materials, e.g. spin glasses [HSS97]. A more recent extension to intermittent behaviour in general can be found in Ref. [FHS08]. Our focus is on the technique used to obtain approximate analytical formulas for the relaxation behaviour.

In this model, nodes stand for regions of configuration space able to support a quasi-equilibrium state. Levels are numbered from zero to $N-1$, and each level comprises the nodes having the same distance to the root node. We assume that every level n node has a degeneracy κ^n. The area of a node stands for its degeneracy, i.e. for the number of microscopic configurations the node contains. The vertical axis stands for the (coarse-grained) potential energy of the configurations belonging to the nodes. The presence of two different energy scales, i.e. L and S corresponding to long and short edges respectively, implies that the nodes at the bottom of the tree have mutually different energies.

The dynamics is given by a master equation, with non-zero rates connecting neighbouring nodes. To fulfil detailed balance, the 'up' rates will be of the form $k_{\text{up}} = \kappa e^{-\Delta/T}$, where Δ is either L or S, while the 'down' rates are all equal to one. The master equation can be dealt with at different levels of approximation. Here we apply the simplest scheme able to capture the essence of the model dynamics, namely the Kramers' approximation scheme earlier discussed. The latter is now applied in an iterative fashion to larger and larger subtrees.

An N level tree consists of two disjoint $N-1$ level subtrees, connected through their common root node. With reference to Fig. (7.1), the leftmost subtree containing the lowest energy configuration, is defined as part 1, and the rightmost as part 2. Each part can be considered as a metastable system. As we will explicitly verify, the time scale for escaping a subtree is longer than the relaxation time scale within the subtree. We are thus poised to apply the Kramers' approximation. Since the latter requires the partition function of each subsystem, let us calculate the partition function Z_N of a tree with N levels, numbered from zero to $N-1$. Quite generally, the partition function is the Laplace transform of the density of states, $\mathcal{D}$ with the inverse temperature $\beta = 1/T$ acting as the Laplace variable, i.e.

$$Z(T) = \int_0^\infty e^{-\beta E}\mathcal{D}(E)dE, \quad \text{where } \beta = \frac{1}{T}.$$

Temporarily assigning energy zero to the top node of the tree, we first note that the nodes of a q level tree emanating from the root have a binomial distribution of energies, since k out of q branches will be of type L, and the rest of type S. Hence the density of the state:

$$\mathcal{D}(E, N) = \sum_{q=0}^{N-1}\sum_{k=0}^{q} \kappa^{N-1-q}\binom{q}{k}\delta(E + (kL + (q-k)S)). \tag{7.9}$$

Performing the Laplace transform and shifting all energy values by $(N-1)L$, an operation which brings the global energy minimum to energy zero, we find:

$$Z(T,N) = (\kappa e^{-L/T})^{N-1} \frac{\rho^N - 1}{\rho - 1} \approx z^{N-1}(T) \quad \text{for } N \gg 1, \qquad (7.10)$$

where

$$z(T) = 1 + \exp(-(L - S)/T), \qquad (7.11)$$

and where

$$\rho(T) = (\exp(S/T) + \exp(L/T))/\kappa = \exp(L/T)z(T)/\kappa. \qquad (7.12)$$

The temperature dependence of ρ and z, the two key parameters in the following development, will mostly be tacitly assumed. We henceforth assume that $\rho \gg 1$, i.e. that the 'up' rates are very small compared to unity, as appropriate at low temperatures. For $i = 1$ or 2, let $Z_i(T, N-1)$ be the partition functions of the two subtrees attached to the root node of an N level tree. Clearly

$$Z_2(T, N-1) = e^{-\frac{L-S}{T}} Z_1(T, N-1).$$

According to Eq. (7.3), the transitions rates describing the probability flow from one part of the tree to the other are

$$k_{21}^{(N)} = v\kappa e^{-L/T} \frac{\rho - 1}{\rho^{N-1} - 1} \approx v\rho^{-N+1} z \quad \text{and} \quad k_{12}^{(N)} = e^{\frac{L-S}{T}} k_{21}^{(N)}. \qquad (7.13)$$

The constant v is henceforth set to one. The relaxation eigenvalue for the N level tree is then

$$\lambda_N = -(k_{12}^{(N)} + k_{21}^{(N)}) = -\rho^{-N+1} z(1 + \exp((L - S)/T)), \qquad (7.14)$$

and the corresponsing relaxation time is

$$\tau_{N,\mathrm{rel}} = -\frac{1}{\lambda_N} = \frac{\rho^{N-1}}{z(1 + \exp((L - S)/T))}. \qquad (7.15)$$

The characteristic escape time out of an N level tree can be obtained in the Kramers' scheme by removing the backflow of probability from the sibling N level tree across the common root node located at level $N+1$. Hence,

$$\tau_{N,\mathrm{esc}} = \frac{1}{k_{21}^{(N+1)}} = \frac{\rho^N}{z}. \tag{7.16}$$

Since $\rho \gg 1$, $\tau_{N,\mathrm{esc}} \gg \tau_N$, as required in the Kramers' approximation.

Consider now the master equation for nearest neighbours hopping on the tree. Its propagator $T_{ij}(t)$, i.e. the probability of being in state i at time t starting from state j at time zero, can be constructed analytically, albeit approximately, using *i)* the sequence $Z_i(T,k)$ of equilibrium partition functions which belong to the nested sequence of subtrees which, starting from a single initial state i, successively reach a state of local thermal equilibrium. And *ii)* the escape times $\tau_{k,\mathrm{esc}.}$ associated to the same sequence. The quantity

$$Q_k(t) = \exp(-t/\tau_{k,\mathrm{esc}}) \tag{7.17}$$

is the probability that at any time shorter than t a trajectory starting at node i has remained within the k level subtree which contains i. We note in passing that $Q_k(t)$ can be evaluated exactly by solving the master equation on the appropriate k level tree, with an absorbing boundary condition imposed at level $k+1$. To avoid overburdening the notation, the dependence of most quantities on the initial state is omitted. The probability that a trajectory has exited the k'th subtree but not the $k+1$'th subtree is $\exp(-t/\tau_{k+1,\mathrm{esc}}) - \exp(-t/\tau_{k,\mathrm{esc}})$. In the Kramers' approximation, this probability redistributes itself 'immediately' according to the equilibrium distribution in the appropriate $k+1$ level tree. Let $d(i,j)$ be the ultrametric distance between the two nodes, i.e. the level index of the root node of the smallest tree containing both nodes i and j. Furthermore, let $P_{k,\mathrm{eq}}$ denote the equilibrium probability distribution in a tree of k levels. Since the latter is at all times mainly concentrated on the low energy states, we assume that the initial state is a local energy minimum. This simplifies the notation with only a small loss of generality. In the special case $i=j$, the ultrametric distance is $d=0$. The probability of staying put at the initial node (a tree with just one level) for a time t is $P_1(t) = \exp(-k_{\mathrm{up}}t) = \exp(-t/\tau_{1,\mathrm{esc}})$, where the last equality defines the escape time from a single node. Accordingly, the Kramers' approximation for the propagator is

$$T_{ij}(t) = \exp\left(-\frac{t}{\tau_{1,\mathrm{esc}}}\right)\delta_{ij} + (1-\delta_{ij})\sum_{k=1}^{\infty}\left(\exp\left(-\frac{t}{\tau_{k+1,\mathrm{esc}}}\right) - \exp\left(-\frac{t}{\tau_{k,\mathrm{esc}}}\right)\right)P_{k+1,\mathrm{eq}}(i). \tag{7.18}$$

As we shall see momentarily, it is possible to write

$$P_{k,\mathrm{eq}}(i) = g_i^k C(i)$$

where $g_j < 1$ depends on the initial state, while $C(i)$ depends on the final state. Using Eqs. (3.40) and (7.16) and reshuffling the indices, we find

$$T_{ij}(t) = \exp\left(-\frac{t}{\tau_{1,\mathrm{esc}}}\right)\delta_{ij} + (1-\delta_{ij})C(i)(1-g_i)\sum_{k=d(i,j)+1}^{\infty}\exp\left(-\frac{tz}{\rho^k}\right)g_j^k - \exp\left(-\frac{tz}{\rho^d}\right)g_j^{d(i,j)+1}. \tag{7.19}$$

We now consider two different initial conditions. If the system is initially located in the highest lying local energy minimum, minimum 1, i.e. the node connected to the root node by a sequence of edges of type S,

$$P_{k,\mathrm{eq}}(i) = \exp\left(-\frac{L-S}{T}(k-1)\right)z(T)^{-k+1}, \quad k = 1,2,\ldots.$$

Hence, $C(i) = 1$ and $g_1 = \exp(-\frac{L-S}{T})/z(T)$. Neglecting the initial decay in Eq. (7.20), and letting $s_l = \rho^l$, where $l = 1,2,\ldots$, we find that rescaling time by s_l leads to a rescaling of the probability:

$$T_{11}(s_l t) = s^{x_1} T_{11}(t), \quad x_1 = \frac{\ln(g_1)}{\ln(\rho)} = -\frac{L - S + T\ln(z)}{L + T\ln(z/\kappa)}. \tag{7.20}$$

Glossing over the fact that the scaling only applies for a countable set of scaling factors, we conclude that the propagator falls off as a power-law with exponent x_1.

Exercise 6: Check the scaling properties given in Eq. (7.20), and discuss the approximations used to reach the result.

Exercise 7: By plotting $T_{11}(t)$ on log-log scales, show that the function displays logarithmic oscillations.

Consider now a system initially placed in the ground state for all subtrees, conveniently dubbed state 2. In this case, $C(2) = 1$ and $g_2 = 1/z(T)$. Retracing our steps, we find

$$T_{22}(st) = s^{x_2} T_{22}(t), \quad x_2 = \frac{\ln(g_2)}{\ln(\rho)} = -\frac{T\ln(z)}{L + T\ln(z/\kappa)}. \tag{7.21}$$

Note that increasing the value of κ, the factor controlling how the degeneracy of the nodes increases with their level, or alternatively, increasing T, the exponent eventually diverges, reflecting the loss of metastability of the tree structure.

Let us finally consider the time dependence of the expectation value of the energy μ_E in an N level tree. Inspecting Eq. (7.18) shows that

$$\mu_E(t) = \exp\left(-\frac{t}{\tau_{1,\text{esc}}}\right) E(j) + \sum_{k=1}^{N} \left(\exp\left(-\frac{tz}{\rho^{k+1}}\right) - \exp\left(-\frac{tz}{\rho^{k}}\right)\right) \mu_{k+1,E}, \tag{7.22}$$

where $\mu_{k,E}$ is the equilibrium value of the energy in the k level tree containing the initial state. Once again, the dependence of the latter quantity on i is left understood. Take now the initial state i as the highest lying minimum. Furthermore, let $\mu^0_{k,E}$ be the thermal energy in a k level tree having $E = 0$ as its global energy minimum, e.g. $\mu^0_{1,\mathrm{E}} = 0$, since a one level tree has only one node. The thermal energy for our sequence of subtrees is then

$$\mu_{k,\mathrm{E}} = \mu^0_{k,\mathrm{E}} + (N-k)(L-S), \quad k = 1,2\ldots N. \tag{7.23}$$

Noting that, for $k > 1$, the temperature and k dependence of $\mu^0_{k,\mathrm{E}}$ is weak, the main contribution to the energy change is the term $(N-k)(L-S)$.

The difference between the two exponential functions in Eq. (7.22), is nearly zero unless $tz/\rho^k \approx 1$. The sum is hence dominated by the term for which $k \approx \ln(t)/\ln(\rho)$. Accordingly the average energy of a system initially placed in the highest lying energy minimum decreases logarithmically as a function of time.

Exercise 8: Calculate $\mu^0_{k,\mathrm{E}}$ from the partition function of a k level tree, and verify that the main contribution to the r.h.s. of Eq. (7.23) stems from the term $(L-S)(N-k)$.

7.3. EXPLORING ENERGY LANDSCAPES

As apparent from the above discussion, potential energy minima, energy barriers and the 'form' of the valleys, the latter captured by the Local Density of States (LDOS) explained in the following, are key quantities for thermally activated relaxation dynamics.

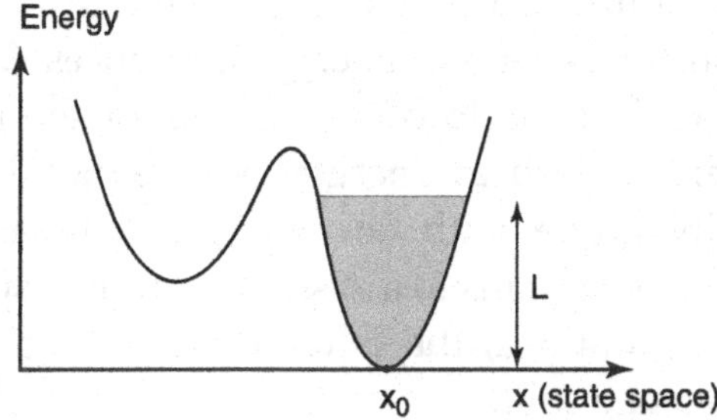

Figure 7.2. The lid method for a bistable potential. Gray area: States counted for lid value L, starting at x_0. The LDOS differs from the true density of states, where everything is included.

Let a catchment basin belonging to a certain energy minimum configuration be the set of all configurations starting from which downhill trajectories ($T = 0$) reach that minimum. An energy landscape can be partitioned into the catchment basins of its energy minima, and a trajectory starting within a basin remains within the same basin for a certain characteristic time, the *escape time* of the basin, much as we saw in connection with the LS tree. Energy minimum configurations are often called *inherent states* [SW83] and can be discovered in a numerical simulation by setting the temperature to zero.

Macroscopic systems have a macroscopic number of configurations and a number of energy minima which, though possibly still large, is usually orders of magnitude smaller than the number of configurations. Much can be gained by coarse graining the full configuration space to the smaller set of local minima, as we did in connection with Example 3. To set up the corresponding coarse-grained dynamics, the calculation and/or estimation of the rates of transition from one minimum to a neighbouring minimum is required. The basic ingredients for a scheme to do so are given by the Kramers' method just discussed. The actual implementation of the scheme requires mapping out all relevant minima and the saddle points which connect them. This highly demanding task has been carried out for a growing number of models of relevance to physics and solid state chemistry. The outcome of the process is a often called *disconnectivity graph* [SPJ96, BK97, Wal03, Ber10]. The latter is plotted as an an upward-rooted tree, where distances along the vertical direction correspond to energy differences. Various energy minima are located at the lower endpoints of the branches, and saddle points are located at energies where two branches split. A disconnectivity graph with few energy minima separated by high energy barriers pertains to a system with correspondingly few meta-stable configurations, the latter representing, e.g., different crystalline phases of a solid.

The connection between the coarse-grained structure of a landscape and relaxation dynamics has been considered by many authors and disconnectivity graphs are routinely used to classify energy landscapes as 'glass formers' or 'structure seekers', the latter being structures which describe, e.g., proteins, materials which are able to quickly find their native state.

If the dominant path between two basins of attraction is the only one of importance, loops can be neglected, and the resulting coarse-grained landscape has then the topology of a tree [HS88]. Importantly, the nodes of such trees represent a number of microscopic configuration, and indeed their degeneracy, which mirrors the so-called Local Density of States (LDOS), plays a large role. Furthermore, landscapes associated with complex dynamics have a very large number of minima, and a broad distribution of energy barriers.

Two related issues are of importance in connection with complex dynamics *i)* the behaviour during the initial quench: Why do glassy systems inevitably end up in states of high energy after a quench? And *ii)* under which conditions is it possible for the system to find a state of near thermal equilibrium which is supported within a small subset of all available states, i.e. a basin of attraction or 'valley' in the landscape? To answer the first question qualitatively, one can imagine ranking all energy minima according to the depth of the corresponding basin, i.e. the magnitude of the smallest barrier which must be surmounted in order to exit the basin. If the frequency of a certain type of minima strongly decreases with increasing depth, a quench will very likely lead to a shallow minimum. An answer to the second question can be obtained numerically using the lid and/or threshold algorithms [Sch10], which both explore the local geometric properties of a catchment basin. Assuming without loss of generality that the energy minimum has energy $E = 0$, the part of the configuration space explored has $E < L$, where the 'lid' or threshold L is a free parameter of the algorithm. The LDOS $\mathcal{D}(\mathcal{E}, \mathcal{L})$ is the density of the microstates at energy E which are found in the basin, or valley, below the given lid. The form of the LDOS controls the nature of thermal relaxation within the valley on time scales shorter than the Arrhenius time associated to the lid, $\tau_A \propto \exp(l/T)$, where, here and henceforth, the Boltzmann constant is omitted for convenience. Numerical studies have shown that the LDOS has a near exponential

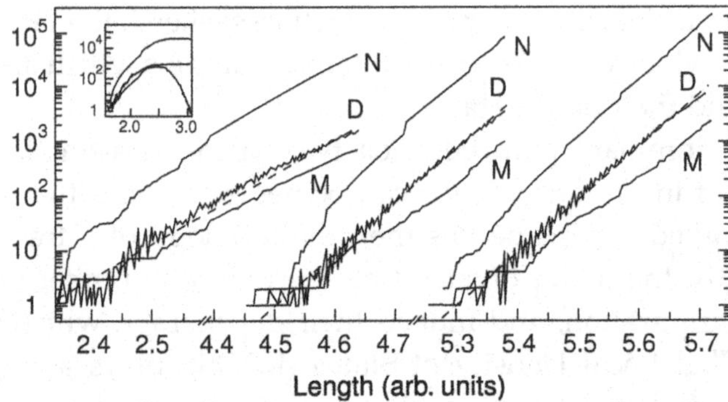

Figure 7.3. The available volume (N), vs the lid, and the local density of states (D) and density of local minima (M) vs the energy for a valley in the energy landscape of the Euclidean TSP. For this problem the energy of a configuration is identical to the length of the corresponding tour. The three sets of data in the main panel are, from left to right, for problem sizes 16, 32 and 36 respectively. The insert shows the same quantities but for an 8 city TSP, where the complete configuration space can be listed. Figure taken from Ref. [SSSA93].

dependence on the energy [SSSA93, SS94, KSH98, BS11a]. Beside the LDOS, Fig. (7.3), taken from Ref. [SSSA93] displays two other quantities characterizing a catchment basin, also called 'pocket' or 'valley', for an Euclidean TSP problem. One is the number N of states below the lid, and the other is the number M of local energy minima below the lid. The near exponential energy dependence of all three quantities underlines the complex relaxation behaviour of the system.

As discussed in Chapter 9, the hierarchical nature of complex relaxation can be inferred from the analysis of experimental data. However, it can also be directly ascertained studying numerical models: Using the lid method, all microscopic configurations of a pocket of a small system can be exhaustively enumerated. The master equation for the system can then be set up and solved numerically, obtaining for each time t the probability density $P(t,x)$ of finding the system in a microstate x, with a sharp initial condition. Two microstates x and y are considered to be locally equilibrated if $\frac{P(t,x)}{P(t,y)} \approx \frac{\exp(E(x)/T)}{\exp(E(y)/T)}$, where near equality is controlled by a small allowed deviation. All microstates can then be grouped into classes.

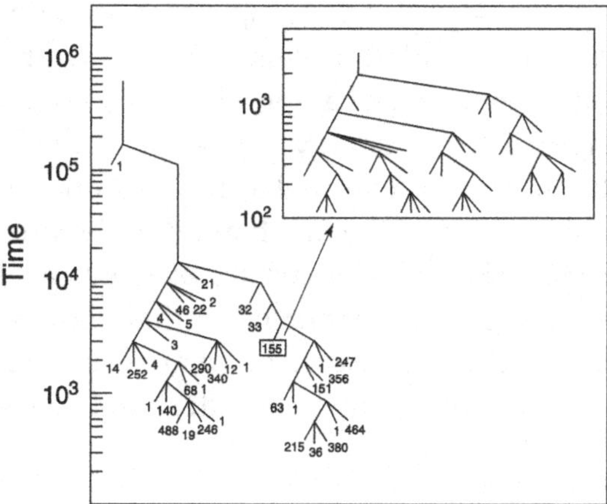

Figure 7.4. The figure, taken from Ref. [SSSA93] shows the hierarchical structure of relaxation at fixed temperature in the state space of a Travelling Salesman Problem (TSP). Each leaf of the tree represents a set of states which are in local equilibrium with one another at a certain level of approximation ϵ. The insert shows how a leaf splits into a whole tree when a smaller value of ϵ is chosen. The vertical axis of the tree is proportional to the logarithm of time: thus the system undergoes a series of local equilibrations (= merging of subtrees) at times almost equally spaced on a logarithmic scale.

Since the system will eventually equilibrate, all states will eventually end up in the same class, and a merging of different classes can be expected during the time evolution of the system. Figure (7.4) shows what happens when the procedure is applied to a small TSP. Equilibration proceeds in stages, with different classes merging, but never splitting again. In the resulting relaxation tree, branches merge at times nearly equally spaced on a logarithmic scale, which, considering the thermally activated nature of the process, indicates that the energies at which different valleys merge are equally spaced on a linear scale.

Let us conclude this chapter with a note on the interplay between real space and configuration space properties: Systems whose degrees of freedom, i.e. particles or spins, are located in a three-dimensional space, possess real space features, e.g. thermalized domains in spin glasses, cages in glass formers, and, in general, spatial correlation functions. Away from any critical points in its phase diagram, i.e. points where spatial correlations are long-ranged, a spatially extended system behaves as an ensemble of independent subsystems each having a linear size of the order of the correlation length. The full energy landscape can then be replaced by an ensemble of similar energy landscapes, one for each of the subsystems. This means on the one hand that the properties of the energy landscapes of small systems are of relevance for extended systems as well, and on the other hand that certain macroscopic dynamical properties can be expected to arise as averages over an ensemble of hierarchical systems, each starting in a different initial condition [KBSR10, SK10].

8

Record Dynamics and Marginal Stability

8.1. INTRODUCTION

As later discussed in Chapter 9, intriguing macroscopic properties, first and foremost the ability to 'remember' any changes incurred by external parameters, are part of the phenomenology of complex dynamical systems. Nevertheless, many complex systems deceptively appear to be in a state of equilibrium when observed on sufficiently short time scales.

To move expediently from short to long time scales is a matter of *coarse-graining* the dynamics. In equilibrium statistical mechanics coarse-graining is initiated by postulating that all microscopic configurations in an isolated system are equiprobable in thermal equilibrium. This very successful axiomatic approach lays the ground for the general applicability of statistical mechanics: Many properties transcend the details of the interactions, precisely because they are statistical in nature. Similarly, complex dynamics has a number of generic properties, i.e. punctuated equilibria in Darwinian evolution resemble intermittency in complex materials. We argue in this chapter that the similarities arise in many cases from the dynamical evolution being driven by record-sized noise fluctuations, events whose temporal distribution is largely independent of the distribution from which the fluctuations are drawn. In the simplified but general coarse-grained description offered by record driven dynamics, for short record dynamics, all important non-equilibrium events are assumed to be triggered by record fluctuations.

To argue for the relevance of the record dynamics scenario, hierarchical energy landscapes are used in this chapter as an initial scaffolding, later relinquished in applications to colloidal dynamics and evolution models, where energy barriers have no dynamical role. Secondly, on the basis of record dynamics we derive expansions for the time dependence of moments, e.g. response and correlation functions, and discuss their applicability to real data. Finally, we compare record dynamics to a different phenomenological description which relies on the concept of effective temperature.

8.2. COARSE-GRAINING COMPLEX DYNAMICS

In thermally activated dynamics, thermal noise, i.e. energy fluctuations due to the interaction with a thermal bath, can induce barrier crossings, i.e. transitions between different metastable valleys. These events set the clock of an *event driven* coarse-grained dynamics, as already seen in connection with the Kramers' approach: The internal dynamics in each valley consists of equilibrium fluctuations exploring a 'local' Boltzmann distribution over the microscopic configuration within the valley. On longer scales, the probability mass is redistributed among different valleys, e.g. a Brownian particle in a sinusoidal potential of amplitude b is confined to a single valley on time scales shorter than $e^{b/T}$ and diffuses, on larger time scales, with an activated diffusion coefficient $D = D_0 e^{b/T}$.

In the (somewhat trivial) diffusion example above, the potential energy is translational invariant in space, and the dynamics is, correspondingly, time translation invariant. Spin glasses and other complex systems are characterized by dilation invariant hierarchical energy landscapes, i.e. nested valleys of the sort epitomized by the LS model shown in Fig. (7.1). Each valley is there represented by a subtree of the full structure, and contains all previously visited valleys as substructures. As discussed in the following exercises, gaining access to a new valley and thereby to an as yet unexplored part of the landscape, does require the occurrence of a record-sized energy fluctuation.

Exercise 1: Consider the random hopping of a point particle on an N level LS tree, where the vertical direction corresponds to the energy. Conventionally, the top node has energy zero. Initially, the particle is located in the highest lying energy minimum, i.e. at the rightmost end node of the graph having energy $-NS$. For convenience we stipulate that $S < L < 2S$. The particle moves as the ball of a pinball machine: Upward moves involve a positive energy change of size ΔE, which is drawn from a fixed distribution $P(x)$. Downward moves involve a series of equiprobable 'right' or 'left' choices, which continue until the particle hits a local energy minimum.
An illustrative, albeit unphysical, choice of $P(x)$ is the box distribution

$$P(x) = 1/b \quad 0 < x \leq b; \quad P(x) = 0 \quad x > b.$$

Plot as a function of b the lowest energy $E_{\text{min}}(b)$ which can be reached by a particle able to overcome a barrier of height b.

Exercise 2: Consider a time series $b_1, b_2 \ldots$ of barrier values and the corresponding time series of lowest attainable energy values $E_{\text{min}}(b_1), E_{\text{min}}(b_2) \ldots$ describing the trajectory of the particle in the above model. Show that a low record in the energy, i.e. an energy value lower than all previously seen implies that a high record in the time series of barrier values has previously been achieved.

Since the distribution of barriers in the LS model is discrete rather than continuous and, furthermore, possesses a lower cut-off, i.e. the smallest barrier has size S, it is clear that not all record-sized fluctuations lead to new non-equilibrium events. As both the discreteness and lower cut-off of the barrier distribution are unimportant for temperatures larger than the smallest barrier size, one is led to consider the limit in which all record-sized fluctuations trigger non-equilibrium events. The idea is rooted in *marginal stability*, a property of the energy landscape of glassy systems, and leads directly to *record dynamics*.

8.3. MARGINAL STABILITY

Broadly speaking, marginal stability refers to how valleys are dynamically *selected* during the off equilibrium stochastic evolution of a complex systems with a large number of metastable valleys. In this section we discuss how marginal stability arises, and why it makes record dynamics possible. Record fluctuations cannot trigger irreversible changes unless a very large number of dynamically inequivalent valleys are available. These valleys must differ with respect to their *dynamical stability*, i.e. their ability to trap trajectories as gauged by different characteristic escape times τ. In thermally activated processes, $\tau = \tau(T)$ is the Arrhenius time corresponding to a free energy barrier $b = Tk_B \ln(\tau)$, where T is the temperature and k_B is the Boltzmann constant.

For record dynamics to be applicable, the distribution of escape times of different valleys must be continuous and strongly biased toward short escape times. Such bias ensures that when the system is quenched, it ends with overwhelming probability in a valley with a very short escape time. The slightest noise fluctuation can then remove it from the initial *marginally stable* or *fickle* state.

At any stage of the subsequent dynamical evolution, all moves leading the system from a more stable to a less stable attractor can be reversed on the time scale at which they first occurred. Such moves correspond to equilibrium like fluctuations within the current attractor. In contrast, moves leading to more stable attractors, henceforth called *quakes*, are irreversible on the time scale at which they occur. Since more stable attractors are increasingly rare, the increase in stability associated to a quake will typically only be marginal. This in turn implies that attaining a record high energy value, measured from the lowest energy of the current valley, is enough to trigger the next quake. At low temperature the (pseudo) equilibrium distribution within the valley is strongly peaked near the lowest energy state. Hence, a random energy fluctuation slightly larger than the previous largest fluctuation, i.e. a record-sized noise fluctuation, can lead to the 'edge' of the valley and trigger a quake. Whence the name *record dynamics*.

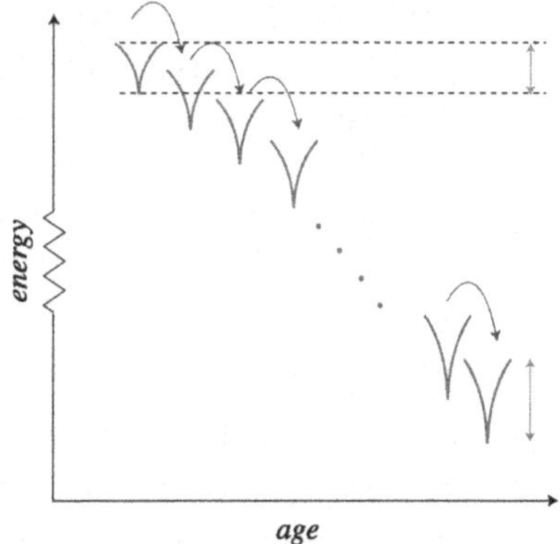

Figure 8.1. The cartoon, taken from Ref. [SD03], illustrates the link between marginal stability and record dynamics: The wedges represent metastable regions, with any reference to their internal hierarchical structure omitted for simplicity. The vertical axis is the energy. Jumps from one metastable region to the next (quakes) are indicated by unidirectional arrows. The distance from the bottom to the upper rim of a wedge indicates its degree of thermal stability.

A cartoon of the physical situation addressed by record dynamics is provided by Fig. (8.1): Each wedge represents an attractor, whose stability is determined by the energy difference between the top and the bottom of the wedge. The stability increases only slightly from one wedge to the next, while the large energy change from the bottom of one wedge to the bottom of the next, de facto renders the corresponding transitions impossible in the direction opposite to the arrows. In the cartoon, the source of irreversibility is a loss of energy to the environment. The same effect can also be achieved by an increase of entropy or by a combination of the two. For graphical clarity, the hierarchical nature of complex dynamics is omitted: A scaled-down size of the whole picture should be iteratively included within each wedge.

8.4. HOMOGENEITY OF RECORD DYNAMICS IN LOGARITHMIC TIME

As discussed in Chapter 5, the temporal statistics of record fluctuations in stationary white noise has universal properties which are unrelated to any physical changes associated with the quakes. Irrespective of the nature the fluctuations, n records occur in the time interval $[t_w, t]$ with probability

$$p_n(t_w, t) = \left(\frac{t}{t_w}\right)^{-\alpha} \frac{(\alpha \ln(t/t_w))^n}{n!}, \tag{8.1}$$

where α is a constant. Whenever all physical changes can be treated as statistically *subordinated* to noise records, a change of independent variable from the (usual) time to its logarithm *trivializes* the dynamics, i.e. it renders the dynamics homogeneous with respect to shifts in the new variable, e.g. on a logarithmic time scale averages change at a constant rate, while on a linear time scale their rate of change is proportional to $1/t$, a clear violation of time translational invariance.

> **Exercise 3:** We argue that ageing dynamics is a homogeneous process when time is replaced by the number of quakes. Is the process also stationary?

Empirical verification

In thermal activated dynamics, barrier crossing is driven by positive energy fluctuations. Once a barrier is crossed and access to a new valley is gained, the energy can decrease noticeably, through an irreversible step, i.e. a quake. Assuming that quakes can be identified from a time series of measurements, one can then check whether their rate decreases as the inverse of the system age in accordance with the prediction of record dynamics. At a more detailed level, one can consider the sequence $t_0, t_1 \ldots t_k \ldots$, where t_k is the time of occurrence of the k'th quake. In a log-Poisson process, the logarithmic waiting times $\tau_k = \ln(t_k) - \ln(t_{k-1})$, $k=1,2\ldots$ are independent and exponentially distributed stochastic variables. Empirically, one can check the extent to which the series $\tau_1, \tau_2 \ldots$ is δ correlated (as a proxy for independence). Secondly, one can check whether the empirical distribution of the logarithmic waiting times is exponential.

Moment expansions

The mathematical tool box widely used for time homogeneous stochastic processes can be adapted to a record dynamics scenario. However, the irreversibility built into the description violates detailed balance, and prevents us from directly using the properties of Markov processes discussed in Chapter 3.

Example 1: Consider the expansion of the propagator, given in Eq.(3.86). Naively, the expression can be rewritten with time replaced by the number of quakes n, and with the explicit dependence on the free energy and temperature written out:

$$T_{ik}(n) = \sum_{\lambda} \left(e^{(F(k)-F(i))/(2T)} \right) \psi_\lambda(i) \psi_\lambda(k) e^{\lambda n}. \tag{8.2}$$

The expression involves a set of eigenstates pertaining to the *coarse-grained* dynamical process which describes jumps between different valleys. The free energy $F(i)$ of valley V_i is given by $F(i) = T \ln Z_i$, where

$$Z_i = \sum_{l \in V_i} e^{-E(l)/T}$$

is the partition function pertaining to the microscopic configurations in valley V_i.
The form used for the propagator relies on detailed balance, a property not conciliable with the irreversibility of record dynamics. Indeed, were the system to revisit the same valley, it would exit it on the same time scale as in the previous visit, and a new record fluctuation would not be required. Clearly, then, record dynamics cannot describe near equilibrium fluctuations.

As illustrated in Fig. (7.4) complex relaxation proceeds through a sequence of partial equilibrations in a hierarchy of nested valleys. Relaxation within each level of the hierarchy has a characteristic time scale, whose inverse is the Perron–Frobenius eigenvalue associated with the restriction of the transition matrix to the valley in question. We can then write the dependence of any moment of the ageing process on the number of quakes as an infinite series of the form

$$M(n) = \sum_{i=0}^{\infty} \exp(\lambda_i n) m_i, \tag{8.3}$$

where the coefficients $m(i)$ and the eigenvalues $\lambda_i < 0$ are unknown and problem dependent. Clearly,

$$\sum_{i=0}^{\infty} m_i = M_0(t_w), \tag{8.4}$$

the initial value of the moment in question when measurements are commenced.

In order to obtain actual time dependences, Eq. (8.2) must be averaged over the distribution given in Eq. (8.1). Everything being linear, it suffices to consider the transformation of a single generic term of the eigenvalue expansion.

Exercise 4: Using Eq. (8.1), show that

$$\sum_{n=0}^{\infty} p_n(t_w, t) e^{\lambda n} = \left(\frac{t}{t_w}\right)^{-\alpha(1-e^{\lambda})}. \tag{8.5}$$

Clearly, the λ 'th term gives rise to a power-law with exponent

$$x(N, \lambda) = -\alpha(N)(1 - e^{\lambda(N)}), \tag{8.6}$$

where we have introduced a possible dependence of both α and λ on the number N of system components. Even though the eigenvectors $\psi_\lambda(i)$ are in general unknown, the propagator itself is a function of the scaling variable t/t_w. Specifically, it is a weighted sum of different powers of that variable. The same is therefore true of its moments, i.e. the magnetization of a spin glass in a field if the initial condition is a delta function, i.e. supported in a single valley. Since one cannot in general solve relaxation problems in large configuration spaces, all different $x(N, \lambda)$ are in principle unknown. Fitting data with the above equation seems then of limited use, given the huge number of free parameters it contains.

Fortunately, the number of terms contributing to the sum is usually very small in large systems. As we shall see, most terms either decay to zero very rapidly, or remain frozen, a behaviour which hinges on the generic N scaling properties of α and λ. The parameter α is proportional to the number of quake events occurring in parallel through the system. Considering that large systems with short-ranged interactions decompose into an extensive number of independent parts, one generally expects

$$\alpha(N) \propto N = \alpha_0 N.$$

From Eq. 8.6, it then follows that $x(N,\lambda) \to -\infty$, unless $\lambda \to 0$ in the limit $n \to \infty$. Clearly all terms for which the last condition is not fulfilled will decay very quickly and can hence be neglected. For the remaining terms the approximation $e^{\lambda} \approx 1 + \lambda$ is valid. Using the expansion, we find

$$x(N,\lambda) = \alpha(N)\lambda(N). \tag{8.7}$$

If λ vanishes faster than $1/N$ in the thermodynamic limit, the exponent $x(N,\lambda)$ goes to zero. The corresponding mode is therefore *frozen*. Important for the dynamics are the modes for which $\lambda \propto 1/N$. Only these modes give rise to finite non-zero exponents which describe measurable changes in a non-equilibrium ageing process where record dynamics applies.

Exercise 5: The correlation function of a stochastic variable F is well suited to describe equilibrium fluctuations. Its behaviour out of equilibrium can however be dominated by a trend, i.e. the time evolution of the average in which case care must be exercised in the interpretation. Consider now an ageing process where the average, $E(M)$, is zero at all times, a criterion satisfied by, e.g., the average magnetization of a spin glass. Then

$$C_M(t,t_w) = \frac{E(M(t)M(t_w))}{E(M^2(t_w))}$$

is the correlation function normalized to one at $t = t_w$. Argue heuristically that, since n, the number of records $n(t,t_w)$, falling between t_w and t, effectively plays the role of time, one can write

$$C_M(t, t_w) = 1 + \sum_{i=0}^{\infty} \exp(\lambda_i n) c_i, \tag{8.8}$$

where

$$\sum_{i=0}^{\infty} c_i = 0. \tag{8.9}$$

Exercise 6: Assume that M in Eq. (8.3) is the linear magnetic response of a spin glass to an external magnetic field, while in Eq. (8.8) the same symbol, used as a subscript right before, denotes the spontaneous magnetization in zero field. Assume that all modes but one, i.e. the mode corresponding to the largest eigenvalue, have decayed to zero in Eq. (8.3) and Eq. (8.8). Find, in this approximation, expressions for the autocorrelation function and the response function. Show then that a mathematical relation between the two ensues, which is similar in structure to the Fluctuation-Dissipation Theorem given by Eq. (3.58).

As described in the above exercise, when a single dominant term describes, asymptotically for large t/t_w, the decay of both correlation and response functions, these two quantities are trivially proportional. This situation appears similar to the fluctuation-dissipation theorem known from stationary processes, and the proportionality can be expressed in terms of an effective temperature [CKP97, BBC03]. However, the mechanisms leading to this FDT-like behaviour are unrelated to equilibrium dynamics, and the effective temperature has no physical interpretation in a record dynamics scenario.

8.5. SCALING BEHAVIOUR IN RECORD DYNAMICS

Physical systems near a second order phase transition have real space configurations which are approximately invariant under a change of length scale. Consequently, such systems feature a time scale invariant dynamics. The physical properties in space and time are then described by power-laws. In contrast, record dynamics usually leads to time dependences which are sums of power-laws with different exponents and which lack scale invariance as long as more than one term in the sum is of importance. However, asymptotically for large values of t/t_w, one term will effectively dominate, and time scale invariance ensues. The latter property is unrelated to length scale invariance in real space. Typically, quakes are real space events with an exponential size distribution, and involve on average a finite number of degrees of freedom. The underlying rescaling symmetry is instead a property of the energy landscape of the systems considered.

A glance at Eq. (8.5) shows that the dynamics has indeed a time scale, albeit a very unusual one, namely the waiting time t_w. The latter is determined by the experimental protocol and is hence not an inherent property of the system itself. A dependence of correlations and other two time averages on the ratio of two times is called *pure* or *full* ageing in the literature. As we shall discuss in connection with applications, systematic deviations from pure ageing behaviour are observed in many physical systems. These can be accounted for by melding record dynamics, inherently based on configuration space properties, with real space descriptions of complex systems.

Part II

Complex Systems with Similar Dynamics

9

Ageing of Spin Glasses

9.1. INTRODUCTION

We have repeatedly mentioned that macroscopic properties of complex systems slowly and systematically change in time. In physics this is called *glassy relaxation,* to indicate kinship to processes in amorphous materials, as well as ageing or *physical ageing,* to indicate that the age of the system since its initial preparation remains imprinted in its state at later times. In contrast, the physical properties of stationary systems are time independent and carry no such information. Experiments and, to a lesser extent, numerical simulations have indeed formed our understanding of ageing phenomenology: Its discovery in *soft* matter [Str78], e.g. polymers, prompted a sustained experimental effort in a class of magnetic materials called spin glasses [LSNB83, AHOR87, Vin91]. The precise measurements of small magnetic effects which are possible in spin glass experiments contributed decisively, and still contributes [RKO03, KRO06, KBSR10], to the development of a multifaceted description of complex dynamics. In this chapter, experimental and simulational results for spin glass ageing are discussed in some detail. Our account focusses on the phenomenology of low temperature dynamics, in particular the analysis of linear response functions and of intermittent fluctuations, which provide direct empirical evidence for our theoretical analysis of complex relaxation.

The brief theoretical discussion at the end of the chapter deals with the static hierarchical configuration space structure peculiar to a mean field spin glass model. The applicability of mean field results to experimental systems with short-ranged interactions has been favoured in some quarters and contested in others. In any case, the conceptual link between the hierarchical structure of configuration space of mean field spin glasses and the dynamical hierarchies discussed in Chapter 6 is tenuous. Nevertheless, static hierarchies can be given a dynamical interpretation with an eye to spin glass applications, and in this context qualitative similarities appear.

Spin glass critical behaviour has been thoroughly investigated experimentally and is crucially important for understanding equilibrium properties, e.g. the nature of the spin glass equilibrium states. It is only mentioned cursorily, since ageing dynamics at low temperatures is only weakly related to equilibrium properties.

The books by Fischer and Hertz [FH91] and Mydosh [Myd93] and the collection of monographs edited by Young [You98] offer a detailed and balanced review of spin glasses and their connection to other complex dynamical systems. This literature should be consulted by readers interested in physics.

9.2. SPIN GLASSES

The term 'spin glass' covers not only a large variety of different *materials*, but also a family of lattice spin models, which, arguably, capture the 'essential' physics of their far more complex experimental counterparts.

Concepts and techniques first developed in spin glasses have migrated to other fields, e.g. neural networks, complex optimization problems and protein folding dynamics to name just a few. Furthermore, a heated debate on the equilibrium properties of spin glasses has influenced how complex dynamics is understood and analysed in the physics community. The discussion below focusses on low temperature ageing and its manifestations, with equilibrium aspects only included to the extent they are relevant to the description of dynamical phenomena.

Spin glasses are characterized by *quenched disorder* and *frustration*: Quenched disorder means that some parameters in the Hamiltonian are random and do not change on all relevant time scales. Frustration [Tou77] means that no microscopic configuration exists which simultaneously minimizes all contributions to the energy function: In the ground state of a spin glass some 'frustrated' interactions always give a positive contribution to the energy.

For concreteness we can think of a spin glass as a set of N spin variables, S_i, residing on magnetic atoms with random positions in a non-magnetic host material. This describes the situation for metallic spin glasses as, e.g., CuMn, where the concentration of magnetic impurities is typically in the range of a few per cent. The fixed positions of the magnetic atoms provide the quenched randomness, while the frustration arises from the sign of the pairwise spin interactions[1] depending on the distance: Some spin pairs prefer a parallel and others an antiparallel alignment.

> **Exercise 1:** The following toy model shows how conflicting requirements produce frustration and ground state degeneracy. Three Ising spins are placed at the vertices of a triangle, each of them interacting antiferromagnetically with its two neighbours, i.e. $J_{ij} = 1$ for $i \neq j$ and i and $j \in \{1, 2, 3\}$. The total energy is the sum of their three pairwise interactions. Show that no configuration exists where all interactions are 'satisfied', i.e. contribute a negative amount to the energy. Find all ground states of this toy model and compare their degeneracy to that of the same model with all interactions equal to -1.

The Edwards–Anderson [EA75] (EA) model is a *bona fide* spin glass system, able to reproduce most experimental features of spin glasses. In this model, Ising spins $\sigma_i, i = 1, 2 \ldots N$ are placed on a regular cubic lattice, usually in two or three spatial dimensions. They interact

[1] For metallic spin glasses this is the RKKY interaction, which is mediated by the conduction electrons of the host metal.

via the first term in the following Hamiltonian

$$\mathcal{H}_{\text{EA}} = \frac{1}{2}\sum_{ij} J_{ij}\sigma_i\sigma_j - H\sum_i \sigma_i. \tag{9.1}$$

The second (Zeeman) term in the Hamiltonian covers the interaction with an external magnetic field H. The bonds $J_{ij} = J_{ji}$ not connecting neighbouring spins are all zero. Other bonds are, for $i < j$, independently drawn from a Gaussian distribution with zero mean and unit variance. The model replaces positional disorder with bond disorder, a simplification which, unfortunately, does not render the model amenable to *ab initio* analytical approaches. In the (vain) hope to obtain a simple analytical description, Sherrington and Kirkpatrick (SK) [SK75] proposed a mean field version of the model, where each spin interacts with all others, and where all real space features are consequently wiped out. The mean field theory of spin glasses turned out not to be as simple as initially surmised.

9.2.1. The Order Parameter

In spin glasses, as in other systems, critical properties and phase diagrams are defined in terms of an *order parameter*, i.e. a non-zero value of the spin glass order parameter defines, by construction, the spin glass phase. However, unlike the situation in a ferromagnet, the magnetization does not qualify as an order parameter in a spin glass, since, in a non-zero magnetic field, it vanishes at *all* temperatures. At sufficiently low temperatures, the i'th spin may well acquire a non-zero thermal average

$$m_i = \langle \sigma_i \rangle \neq 0. \tag{9.2}$$

However, averages belonging to different spins will point in random directions, and their sum, i.e. the total magnetization, will therefore lack extensivity in the thermodynamic limit. A better choice of order parameter is the Edwards–Anderson order parameter [EA75]:

$$q_{\text{EA}} = N^{-1}\sum_i m_i^2 \tag{9.3}$$

which, being a sum of squares, is non-zero as soon as at least some of the m_i acquire a non-zero value.

> **Exercise 2:** Argue that the global inversion symmetry of the spin glass Hamiltonian is broken if the Edwards–Anderson parameter is non-zero. In other words, a canonical average would produce a zero order parameter at all temperatures.

The EA order parameter can also formally be expressed as the infinite time limit of the spin-spin autocorrelation function:

$$q_{\text{EA}} = \lim_{t\to\infty} N^{-1} \sum_i \langle \sigma_i(t)\sigma_i(0)\rangle = N^{-1} \sum_i m_i^2. \qquad (9.4)$$

It should however be kept in mind that the limit is experimentally and numerically very hard to reach.

9.2.2. Are Spin Glasses in Equilibrium?

Early in the history of spin glasses, magnetic noise measurements were instrumental in elucidating why these systems, under certain conditions, deceptively appear to be in thermal equilibrium: Using an oscillating (AC) field as a probe produces an oscillating response function which depends on both the frequency ω of the applied field and, as it eventually became clear, on the age t_w at which the field is applied. Comparing the imaginary part of the AC susceptibility, χ'', to the power spectrum $S(\omega)$ of the spontaneous magnetic fluctuations, see Eq. (3.65), reveals any deviations from the Fluctuation-Dissipation Theorem (FDT).

Reim *et al.* [RKM⁺86] measured both the frequency dependent susceptibility and the magnetic noise autocorrelation function for $Eu_{0.4}Sr_{0.6}S$ as a function of temperature. Using 50 and 10 Hz data, they concluded that the Fluctuation-Dissipation Theorem is fulfilled at all temperatures, and interpreted their findings as a sign that the sample was thermally equilibrated, even though they acknowledged that the system might be non-ergodic. Refrigier *et al.* [ROB87] measured the equilibrium magnetic noise of the insulating spin glasses CsNiFeF6 and CdIn0.3Cr1.7S4 at various temperatures above and below the transition temperature and for frequencies in the range $10^{-2} - 10^{3}$ Hz. They came to the same conclusions as Reim *et al.* regarding the FDT and were also able to determine the imaginary part of the equilibrium susceptibility as a function of temperature. The growing realization [LNSB85, NSLS86] that the time elapsed since the thermal quench affects the measurements and that spin glasses are not really in equilibrium is clearly summarized in the words of Alba *et al.* [AHOR87]:

> *"We have here evidence that the spin glass does not reach its equilibrium state. Nevertheless, it has been evidenced that at observation times much shorter than the time spent at constant temperature before any magnetic field change, the relaxation is representative of the dynamics at thermodynamic equilibrium."*

The term 'equilibrium behaviour' will be used in this chapter as a synonym for the (longish) 'pseudo-equilibrium behaviour displayed on sufficiently short observational time scales'.

Summarizing, in spin glasses and other glassy systems, the FDT is fulfilled at high frequencies $1/\omega \ll t_{\mathrm{w}}$, i.e. short probing time scales. For longer time scales, $1/\omega \gg t_{\mathrm{w}}$ the FDT is violated. For $t_{\mathrm{obs}} = 1/\omega \ll t_{\mathrm{w}}$, the system appears to be in a state of local thermal equilibrium. The situation can be visualized as entrenchment in valleys of the energy landscape whose escape time grows linearly with t_{w}.

Order parameter susceptibility and AC susceptibility

There is a general consensus [You98] that three-dimensional spin glasses in a zero external magnetic field have a second order phase transition: Below a temperature T_c, which depending on the type of spin glass is of the order of a few tens of degree Kelvin, $q_{\mathrm{EA}} \neq 0$. Since in zero field the magnetization of a spin glass vanishes at all temperatures, the linear magnetic equilibrium susceptibility has the Curie temperature dependence earlier given in Eq. (3.55):

$$\chi_{\mathrm{M,eq}} = \frac{\sigma^2_{\mathrm{M,eq}}}{k_B T}. \tag{9.5}$$

The function has no singularities for $T > 0$. By way of contrast, the spin glass susceptibility (or order parameter susceptibility) diverges at the critical point. Following Fischer and Hertz [FH91], the latter is given by:

$$\chi_{SG} = \overline{\left(\frac{\langle \sigma_i \sigma_j \rangle - \langle \sigma_i \rangle \langle \sigma_j \rangle}{k_B T} \right)^2}, \tag{9.6}$$

where the overbar denotes an average over all sites. The existence of a spin glass phase in zero magnetic field has been shown experimentally and numerically. In a non-zero magnetic field its existence seems ruled out by both experiments [MJN+95] and numerical model simulations [JKK08]. However, the issue somehow remains theoretically contentious.

The (static) linear magnetic susceptibility given in Eq. (9.5) can be expressed as the limit $\omega \to 0$ of the frequency dependent AC susceptibility $\chi_{AC}(\omega)$. The latter is the response to a harmonically oscillating field of unit amplitude. The field is switched on at time t_w after the quench, and depends both on ω and on t_w. Figure (9.1)

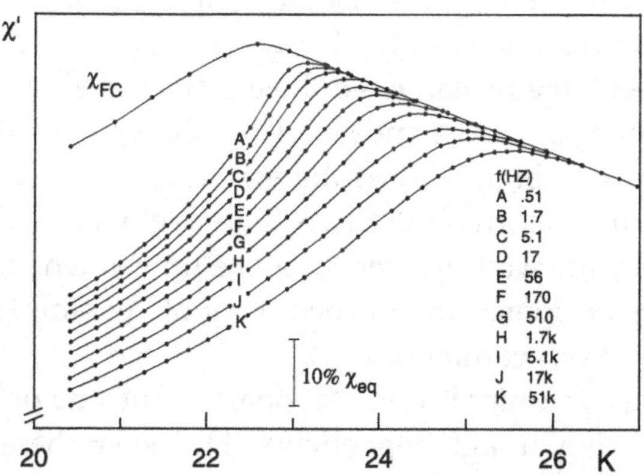

Figure 9.1. The uppermost curve is the field cooled susceptibility, while the other curves are zero field cooled susceptibilities for a set of frequencies equally spaced on a logarithmic scale. All data are plotted vs the temperature. The reciprocal of the frequency at which the AC susceptibility data start deviating from the field cooled data provides an estimate of the relaxation time in the system. Data redrawn with permission from Svedlindh *et al.* [SGN+87].

describe the temperature and frequency dependence of the real part of the AC susceptibility. As the temperature decreases, the susceptibility increases until a cusp is reached, whose position moves to the left in a nearly logarithmic fashion. The top curve shown is the field cooled (FC) susceptibility χ_{FC}, which is obtained by imposing a field before the thermal quench. Note that the limiting curve one would obtain by extrapolating to $\omega = 0$ is very different from the FC susceptibility. The difference is called remanence, and its presence clearly demonstrates that spin glass relaxation is a non-equilibrium phenomenon.

9.2.3. Ageing: Experimental and Numerical Observations

Ageing experiments require preparing the system in a non-equilibrium state, e.g. via a thermal quench which brings it from a state of thermal equilibrium at a high temperature to a glassy state at a lower temperature. The equivalent step for colloids consists of centrifuging the system to a desired (high) density. While quenches can never happen instantaneously, except in computer simulation where one simply changes the value of a parameter, many aspects of ageing can nevertheless be modelled by assuming that the quench is instantaneous. This allows one to define the system age t unambiguously as the time elapsed since the quench. In any experiments involving two time averages, a second time, conventionally the waiting time t_w, is the time at which an external field eliciting a response is switched on or off, or simply the time at which observations commence.

As ageing dynamics is non-stationary, some one point averages must exist having an age dependence. However, these might well be impossible to measure. Conversely, more easily measured one point averages can be age independent and hence only say little about the ageing process. Consider, e.g., a spin glass: As numerical model simulations have shown, the energy of a spin glass model following a quench is far too high, compared to its thermodynamic equilibrium value at the ageing temperature. In the course of the ageing process, the (model) energy slowly decays. The effect would be very hard to notice in an experimental setting, as magnetic interactions are very small compared to other non-magnetic contributions to the energy, e.g. lattice vibrations. Conversely, the spontaneous magnetization is more readily available, but remains on average zero in a spin glass systems. There are two ways around this problem: Historically, linear response functions, e.g. the magnetization elicited by switching a small external field on or off at a certain point of time have played a large role. Response and correlation functions are two point averages, and in a non-stationary situation depend on two time arguments. This fact has been used to identify the presence of ageing and has even been used as its defining property. The latter definition is unnecessarily reductive, since, in some cases, e.g. in dense colloids, ageing phenomena are visible without any externally imposed perturbation. Nevertheless, the importance of linear response experiments for mapping out ageing phenomena, including so-called memory and rejuvenation phenomena, can hardly be overstated. A second and more recently introduced probe of ageing behaviour are fluctuation spectra, which, in ageing systems, display a characteristic intermittent behaviour.

Linear response

A well established protocol for spin glass ageing experiments is the Zero Field Cooled (ZFC) protocol: The system is quenched below the glass temperature T_g in zero external field, and a small magnetic field of constant magnitude is imposed at a later time t_w. The linear magnetic response is then measured as a function of time t, and more often as a function of the observation time $t_{obs} = t - t_w$. Tradition dictates that the observation time be called t, and that, effectively, the time axis for the measurements begin when the field is switched on. This switch is closer to an instantaneous event than the initial quench is. From a theoretical point of view, using the end of the initial quench as the origin of the time axis nevertheless presents a considerable advantage.

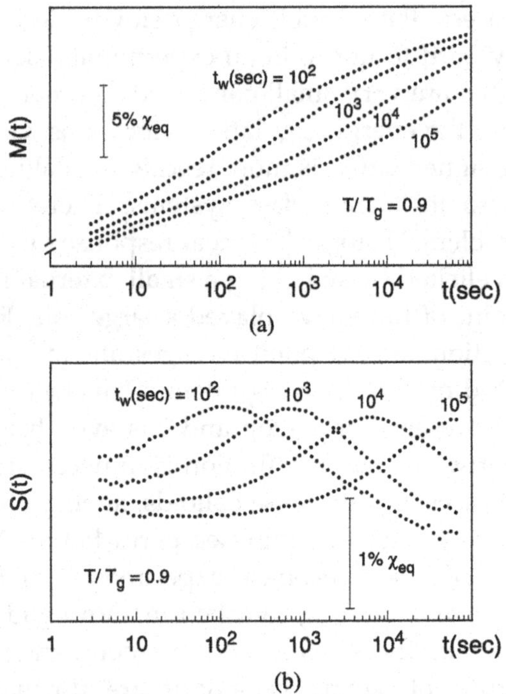

Figure 9.2. The upper panel is the ZFC magnetization of a spin glass for different ageing times, as a function of observation time. The lower pane shows as a function of the observation time the logarithmic derivative of the same quantities. Note that the age of the samples t_W can be read off as the location of the maxima. Adapted with permission from Svedlindh *et al.* [SGN+87].

All data shown in Fig. (9.2) are adapted with permission Svedlindh *et al.* [SGN+87]. As anticipated, time t (i.e. $t_{\rm obs}$ in our notation) is measured from the instant at which the field is switched on. The upper panel displays the magnetization $M(t_{\rm obs})$, and the lower panel its logarithmic derivative $S(t_{\rm obs}) = \frac{dM}{d\log(t_{\rm obs})}$. Clearly, the data show a systematic dependence on t_W: The response decreases the older the system is. Secondly, an embryonic memory effect can already be detected: It is possible to reconstruct from the shape of the magnetization the time at which the field was switched on. This time approximately corresponds to the location of the maximum of S. Clearly then, M and S are functions of two variables, namely

t_{obs} and t_w. It is however possible to approximately express M as a function of the single scaling variable t/t_{obs}.

In Thermoremanent Magnetization (TRM) experiments, the system is first magnetized at a high temperature by a small applied DC magnetic field and then, at $t = 0$, cooled down to a temperature $T < T_g$. At time $t = t_w$ the external field is cut, and the ensuing decay of the magnetization $M_{TRM}(t, t_w)$ is observed. To see why the ZFC and TRM protocols give equivalent information on the ageing dynamics as far as *linear* effects are concerned, we now consider the combined effect of two perturbing fields of equal magnitude. The first is turned on (the ZFC case) and the second turned off (the TRM case) at time $t = t_w$. Since the resulting field, which is the sum of the two, is independent of time, the corresponding response is the so-called field cooled magnetization M_{FC}. By the superposition principle we thus have, for all t_w's:

$$M_{FC}(t) = M_{ZFC}(t, t_w) + M_{TRM}(t, t_w). \tag{9.7}$$

Exercise 3: Why is the Field Cooled Magnetization independent of t_w? For an empirical demonstration of Eq. 9.7 see, e.g., Ref. [Vin91].

The systematic dependence of the magnetization curves on the waiting time t_{obs} suggests that they could be functions of a single *scaling variable*, e.g. t/t_{obs}.

The proper scaling of response functions has been an open issue for many years, with the first analysis performed by Alba *et al.* [AOH86], who analysed the TRM curves for the spin glass Ag: Mn(2.6%). An important development in this area was the discovery of the 'end of ageing' effect, which is shown in Fig. (9.3): The upper panel shows a set of TRM decay curves corresponding to different *effective* waiting times t_w, which are estimated from the position of the maximum in the $S(t)$ curves shown in the upper corner of the figure. The term 'effective' covers the fact that even in the case when the field is switched on immediately after the end of the cooling process, the $S(t)$ curves maintain their characteristic maximum at a small yet non-zero value of t. Clearly then, identifying precisely the

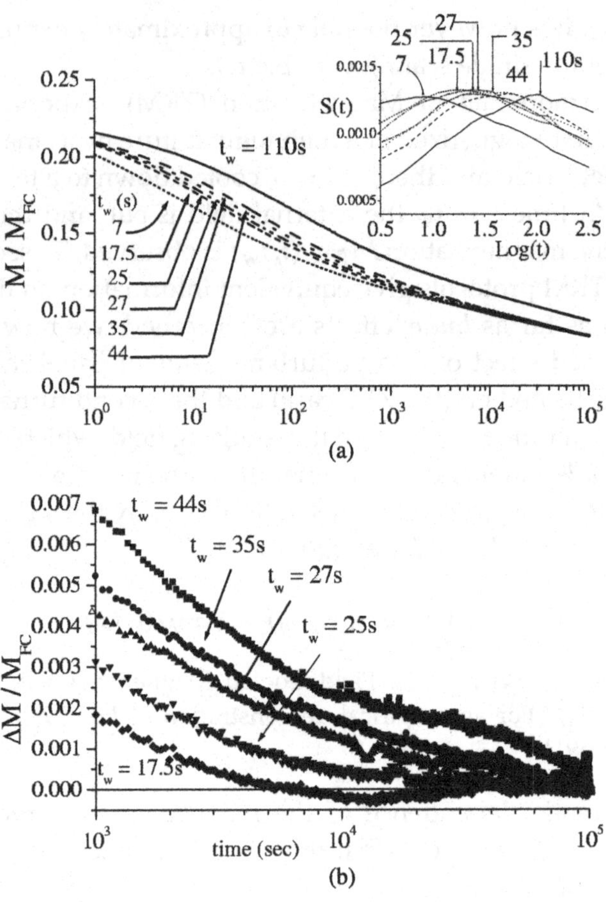

Figure 9.3. Experimental results by Kenning *et al.* [KRO06], illustrating the loss of t_W dependence of the TRM which appears for very long observation times.

end of the initial quench and, correspondingly, the beginning of the isothermal ageing process, is a difficult task. The lower panel of the figure shows, as a function of the observation time t_{obs}, a blow-up of the time dependence of five TRM curves corresponding to different values of t_W. In these curves a fast initial decay of the TRM has been subtracted. Initially, as the curves are nearly parallel, they could be superimposed by using a suitable scaling variable as abscissa, e.g. naively, t/t_W. However, since the curves merge at large values of the observation time, all t_W dependence is eventually lost. In light of this result, any scaling form, including the widely used *sub-ageing*, i.e. t/t_{w}^{μ} scaling, with $0 < \mu < 1$, is an empirical approximation with a restricted range of validity. To explain the end of ageing behaviour one needs to combine hierarchical and real space features of spin glass relaxation [KBSR10, SK10], a topic which falls outside the scope of our present discussion.

Memory and rejuvenation

Combining ageing with temperature changes in magnetic AC susceptibility measurements demonstrates the so-called memory and rejuvenation effects [JVH+98] which are generic features of complex dynamics. Starting from a high value, the temperature is decreased at a constant rate, down to a certain value $T_{\mathrm{stop}} < T_{\mathrm{g}}$. The frequency of the external magnetic field is sufficiently high to ensure that the curve obtained during the cooling process is equilibrium like. At $T = T_{\mathrm{stop}}$, cooling is discontinued and the system is allowed to age for a long time, i.e. for several hours. During the ageing process, the susceptibility decreases, leading to the 'dip' shown in Fig. (9.4). Interestingly, when the cooling is resumed, the susceptibility curve quickly returns to its higher equilibrium shape (i.e. the shape which can be measured without ageing stops). In other words, the time spent ageing at a certain temperature has no consequences at lower

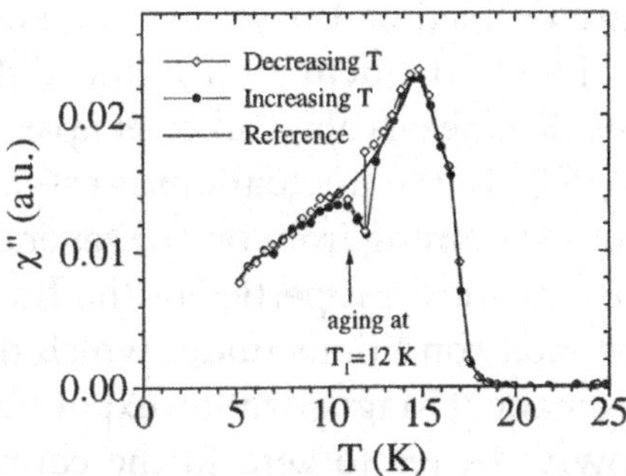

Figure 9.4. Experimental results by Jonason *et al.* [JVH+98], illustrating the fact that the memory of the configurations visited by ageing at a certain temperature is not erased by a temperature sweep.

temperatures, where the system then looks younger. This amounts to a *rejuvenation* effect upon cooling. Upon reheating the system at *constant* rate, i.e. without any stops, the dip in the susceptibility curve near $T = T_{\text{stop}}$ is reproduced. This is amounts to a *memory* effect upon reheating, since the thermal protocol to which the system was subjected can be dynamically recalled at later times.

Intermittency

Intermittency is, among other things, a property of fluctuation spectra of ageing systems. Since the analysis of such spectra does not require the ability to impose and control an external field, this type of analysis can be carried out on purely observational data. Furthermore, the dichotomy between equilibrium and non-equilibrium dynamics, which was first demonstrated in linear response experiments, manifests itself in a particularly clear fashion in this approach. The techniques needed to perform intermittency analysis are simple and general, and yet not widely known. This motivates our detailed discussion below.

The starting point is always a time series $M(t_i), i = 0, 1, 2, \ldots$ comprising the observed values of the macroscopic quantity of interest, e.g. the magnetization M. Measurements are taken at equidistant times and spaced by the sampling interval $\delta t = t_i - t_{i-1}, i = 1, 2, \ldots$. By definition of t_w, the first measurement is taken at age $t_0 = t_w$.

Fluctuations are defined as the set of differences $\delta M_i = M_i - M_{i-1}, i = 1, 2 \ldots$, i.e. each fluctuation is the difference between two values of a physical observable, taken δt apart. The Probability Density Function (PDF) of the fluctuations is estimated empirically using the δM_i values stemming from one or several time series.

Obviously, the statistical properties of the time series at hand reflect the experimental conditions under which the data are measured. In Thermoremanent Magnetization experiments, the average magnetization slowly decays to zero in the course of the ageing process. In contrast, if the magnetic field is zero, the (spontaneous) fluctuations have zero average at all times. Properties of fluctuation time series taken under otherwise identical conditions depend both on the time window of observation and on the sampling time δt. Ideally, the length of the series, i.e. the total observation time should be rather short, with good statistics provided by considering a large ensemble of independent time series. With a short series, only little ageing occurs during the observation process, and any properties extracted pertain to the system of age t_W. Choosing an observation time of the order of t_W, i.e. taking observations between times $t = t_W$ and $t = 3t_W$ is a reasonable compromise between statistical accuracy and time definition. This choice leaves us with two free parameters, t_W and δt. Clearly, the latter can be varied by appropriately pruning the time series, e.g. skipping every second datapoint.

Exercise 4: A time series consisting of Thermoremanent Magnetization (TRM) values has a time dependent average, and is clearly not stationary. Conversely, spontaneous fluctuations in zero field have zero average. Does this mean that the corresponding series is stationary? If not, which statistical properties of the series could depend on time?

The uppermost plot in Fig. (9.5) describes numerical results stemming from simulations of the EA model (see Ref. [Sib07] and Eq. (9.1)) which were performed under the ZFC protocol at temperature $T = 0.3$. After an instantaneous quench in zero field, a small field, $H = 0.1$, is switched on at age $t_W = 1000$, the same time at which observations commence. Since the vertical axis is logarithmic, the plotted PDF of the magnetization fluctuations has a zero-centered Gaussian part, which appears as a parabola, flanked by an exponential tail, the straight line. The exponential tail seen on the right flank of the PDF represents intermittent fluctuations. These are large and rare fluctuations,[2] which in the present example are irreversible, and hence determine the build-up of a positive average magnetization in response to the applied field. The lower plot in Fig. (9.5) describes the energy fluctuations of the EA model. We note in passing that a similar plot would be hard to obtain based

[2]Still, they are much more frequent than hypothetical fluctuations of the same size, but drawn from the Gaussian distribution describing the central part of the data.

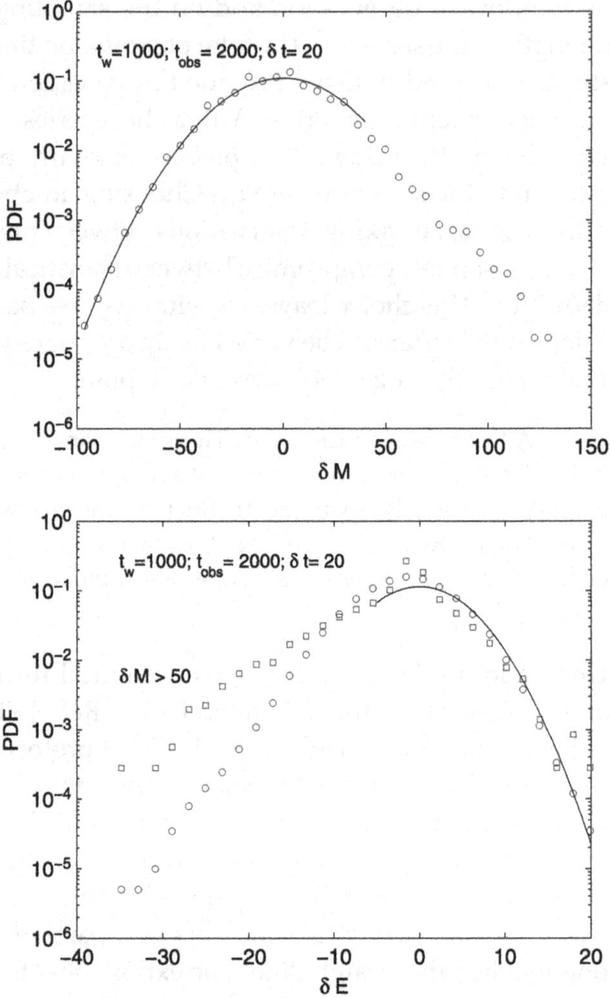

Figure 9.5. Top: the PDF of the ZFC magnetization fluctuations in a field H = 0.1, with $\delta t = 20$, $T = 0.3$ and observation interval [1000, 3000]. The statistics is based on 2000 trajectories. The line is a zero average Gaussian, fitted in the interval $-80 < \delta M < 50$. Bottom: PDF of the energy fluctuations (circles), and a Gaussian fit to positive fluctuations (line). The squares describe energy fluctuations falling in the immediate vicinity of large ($\delta M > 50$) magnetization changes.

on experimental spin glass data, as the magnetic contribution to the energy is dwarfed by other contributions, e.g. lattice vibrations. The lower panel of the figure depicts the energy fluctuation PDFs of two sets of data obtained for the same parameter values as the upper panel. The two data sets coincide for positive values of the abscissa, where the shape of the PDF is Gaussian, but split on the left flank. Again, the Gaussian fluctuations are reversible and the intermittent ones, our *quakes*, are irreversible. It is clear then, that in thermal activated ageing, the non-equilibrium part of the dynamics is intimately connected to the quakes.

Of the two data sets shown, the one where the intermittency is weaker describes the conditional PDF obtained by excluding from the statistics all quakes not occurring in sync with large magnetic fluctuations. Since the statistical weight of the tail is more than tenfold smaller than in the first PDF, which describes all fluctuations, large positive magnetic fluctuations and large negative energy fluctuations are strongly correlated in time.

Exercise 5: Predict or calculate the appearance of the PDF of the x-coordinate fluctuations of a Brownian particle, see Eq. (2.32).

Exercise 6: Same question as in the exercise right above, but for a positively charged Brownian particle moving in an electric field pointing in the positive x direction. Do you find any important difference between the PDF of Brownian fluctuations, with or without a field, and the PDF's depicted in Fig. (9.5) ?

Classifying, within a trajectory, a single fluctuation as being either reversible and equilibrium like or as an irreversible event requires a choice of threshold which is, to a degree, arbitrary. Yet, the picture emerging from the statistics is clear: The overwhelming majority of the fluctuations are reversible, i.e. equilibrium like, and occur around a plateau value, characteristic of the 'current' catchment basin or metastable valley of the energy landscape of the system. The tails reflect occasional transitions from one basin into the next. The (generally observed) fact that the size distribution of intermittent events is exponential, indicates that such transitions only involve a limited number of degrees of freedom, e.g., for the EA model they correspond to flipping a small cluster of contiguous spins, i.e. a real space domain. We are now ready to discuss how the values of δt and t_w affect the shape of the fluctuation PDF. Since the Gaussian part of the PDF has zero average, it is fully characterized by its width. We refer to the square of the width as the Gaussian variance σ_G^2, not to be confused with the true variance of the PDF, which is indeed mainly determined by the intermittent tail or tails. As discussed in the exercise below, whenever δt exceeds a characteristic time τ, one finds

$$\sigma_G^2 = 2\sigma_{M,eq.}^2 \tag{9.8}$$

where $\sigma_{M,eq.}^2$ is the equilibrium variance of M. Consequently, the Gaussian distribution of quasi-equilibrium fluctuation is independent of δt, for sufficiently large δt. A dependence on t_w would imply that different valleys support different types of equilibrium fluctuations. This is possible but, to the best of the authors' knowledge, has not yet been observed.

Exercise 7: Identify τ as a property of the (pseudo) equilibrium autocorrelation function of M, and argue that Eq. (9.8) is correct.

Glossing over the fact that a knife sharp distinction between Gaussian and intermittent fluctuations cannot be strictly maintained, we note that intermittent fluctuations being very rare, the chance that more than one will fall in the same δt can be ignored.

The area below the intermittent wings of the (area normalized) PDF is then precisely the probability that an intermittent fluctuation, regardless of its size, fall in an interval δt. This probability vanishes as $\delta t \to 0$ and, for sufficiently small δt, is proportional to δt. Considering that the Gaussian part of the PDF is not effected by the value of δt, the relative weight of the intermittent tail (or tails) increases, for fixed t_w, linearly with δt.

In thermally activated ageing, intermittent events are connected to moves from one attractor basin to another in the energy landscape, and their probability depends in an Arrhenius fashion on the height of the energy barrier b separating these basins, i.e. it is proportional to $\exp(-b/T)$, where T is the temperature. If, as expected, the relevant barriers increase during the ageing process, the weight of the intermittent tail correspondingly decreases. Thus, in a well aged system, only equilibrium like fluctuations will be discernible and the PDF will appear to be purely Gaussian. A scaling analysis of intermittency data showing how the intermittent tail in the fluctuation PDF disappears with increasing t_W sheds light on the height of the typical energy barriers which is surmounted at age t_W.

Figure (9.6) is concerned with the scaling properties of the energy fluctuations of the EA model, and shows several PDF's obtained for pairs of values of δt and t_W with $\delta t = 0.01 t_W$. Their collapse

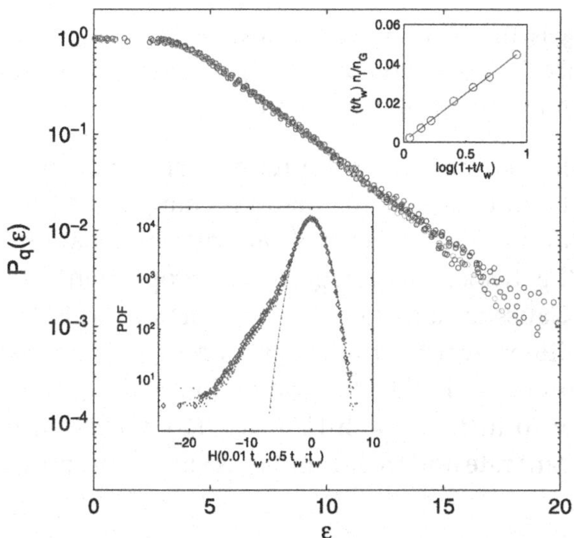

Figure 9.6. The probability $P_q(\epsilon)$ that the amount of energy intermittently released is larger than ϵ, is plotted versus ϵ for ages $t_W = 1000, 2000, 4000, \ldots 64000$. Lower insert: Unnormalized PDF of the heat transfer $H(0.01 t_W, t_W/2, t_W)$. The full lines, indistinguishable on the scale of the plot, are fits to Gaussians with zero average. Upper insert: The relative weight nI/nG (see text) of the intermittent tail is obtained varying t for fixed δt from seven values of $H(0.005 t_W, t, t_W)$. Here, $t_W = 4000$ and $t_{obs} = 200, 600, 1000, 2000, 3000, 4000$ and 6000. The straight line is predicted by the theory. The simulation temperature is $T = 0.3$. Figure taken from Ref. [SJ05].

indicates that the rate of intermittent events decays as the inverse of the system age, a property shared by several complex dynamical systems.

Exercise 8: Assume that the (Arrhenius) rate r_A of intermittent events in thermally activated ageing at temperature T depends on the age t as

$$r_A = \frac{c}{t}, \tag{9.9}$$

where c is a dimensionless constant. Let the energy barriers surmounted at age t have a size $b(t)$. Show that

$$b(t) = T \ln\left(\frac{t}{t_0}\right), \quad t > t_0, \tag{9.10}$$

where t_0 is the time associated with the crossing of a barrier of height $\approx T$. The condition $t > t_0$ is physically reasonable in connection with intermittency. Why?

Let us finally estimate how the relative statistical weights of the intermittent and Gaussian fluctuations change as a function of the observation time. The number of observed fluctuations n_G and n_I falling in the respective categories is proportional to the areas below the Gaussian and intermittent part of the PDF. Assuming that intermittent events at any age t occur at a decelerating rate $1/t$, their number n_I in the observation interval (t_w, t) is on average proportional to $\ln(t/t_w) = \ln(1 + t_{obs}/t_w)$. Gaussian fluctuations have a constant rate and their number is simply proportional to t_{obs}. Hence,

$$t_{obs}\frac{N_I}{N_G} \propto \ln(1 + t_{obs}/t_w). \tag{9.11}$$

The upper insert in Fig. (9.6) shows that the relation is empirically satisfied by the EA simulation data. We note that t in the plot corresponds to our present t_{obs} and that the ordinate is, misleadingly, also scaled by $1/t_w$. This does not affect the plot, since while t_{obs} is varied, $t_w = 4000$ remains constant.

9.3. THEORETICAL CONSIDERATIONS

Changes of catchment basin or valley are the salient events which appear as intermittent fluctuations in fluctuation spectra taken from ageing systems. As the rate of these events continuously decreases with the system age, the height of the barrier leading from the current valley to the next, yet unexplored, valley continuously increases. Figure (8.1) is a cartoon emphasizing the slow growth of barriers but neglecting any barrier structure within each valley. Considering that initial quenches generally lead into a shallow attractor suggests that shallow attractors and correspondingly small barriers are pervasive in the energy landscape of spin glasses and other glassy systems. A deep valley is thus likely decorated with smaller valleys, not unlike the multi scale structure of a real mountain landscape, where the size of the peaks ranges from pebbles to Mount Everest.

The relaxation tree depicted in Fig. (7.4) for TSP relaxation, and a similar tree obtained for the EA spin glass model [SS94] show the hierarchical nature of thermal equilibration processes in these two cases and point to a hierarchically organized coarse-grained picture of the energy landscape: A tree whose nodes represent metastable regions which either contain a local energy minimum or a saddle point [HS88, SH89, SH91]. Finally, Fig. (9.4) [JVH+98] provides important experimental evidence pointing in the same direction: As already discussed, cooling the system after ageing isothermally at the temperature of the 'dip' in the susceptibility data, brings the susceptibility back to the (pseudo) equilibrium master curve. The effect of isothermal cooling diappears and the system is thus rejuvenated by cooling. Since as also shown in Ref. [JVH+98] additional dips can be created by a new ageing process starting at low temperatures, cooling at any T is akin to the quench which starts the ageing process itself: It creates a random condition in the part of configuration space delimited by barriers of order T, i.e. the part dynamically accessible at the relevant temperature. This implies the presence of a wide distribution of barrier sizes. However, the 'smoking gun' with regard to the hierarchical organization of the spin glass landscape is that upon reheating at a constant rate, the previously created dip reappears. The system apparently returns to the part of configuration space explored in the ageing process, and the finer barrier structure explored at the lower temperature is hence embedded in the latter. Importantly, memory and rejuvenation effects of the sort just discussed are widespread in complex systems, e.g. they are found in ferroelectric materials [CCF+04, CCFG10] and gelatine gels [NMRP00]. Following the seminal ideas of Simon [Sim62] already discussed in Chapter 6, hierarchies of time scales are a natural setting for complex dynamics and their presence underpins many of its general properties. In the physics community, the route most commonly used to reach a hierarchical picture of configuration space is based on the equilibrium properties of the Sherrington and Kirkpatrick model, henceforth SK model, described by Parisi [Par79, Par80, Par83]. Glossing over a 30 year long dispute on the virtues of the SK model as a proxy for real spin glasses with short range interactions, the approach has the undesirable, if unintended, consequence that any hierarchical feature seen in experiments or numerical simulation is routinely interpreted as corroborating the validity of the SK model and more generally of mean field models as paradigms of complex dynamics. The *ultrametric structure* which emerges from the Parisi solution of the SK model and its dynamical interpretation are discussed below.

Equilibrium states in the SK model

Sherrington and Kirkpatrick [SK75] considered the EA model with Gaussian distribution of interactions with mean $J_0 \leq 0$, rather than zero (a positive J_0 corresponds to an average ferromagnetic interaction among the spins). To perform the quenched average SK used the now famous 'replica trick'.[3] The first step is to note that

$$\log(Z) = \lim_{n \to 0} \frac{Z^n - 1}{n}. \tag{9.12}$$

[3]SK credit M. Kac for inventing this device. A very similar procedure is however also utilized by EA [EA75], who did not use the name 'replica trick'.

When n is integer, Z^n is just the partition function of n independent, non-interacting systems, the so-called replicas. Performing the quenched average over the site dependent interactions replaces them with interactions among the replicas. The free energy of the system then takes the form of a multidimensional integral over replica indices, which SK solved by applying the steepest descent method and by assuming that all replicas are equivalent (the replica symmetry assumption). By the analytic continuation $n \to 0$, SK finally obtain coupled formulae for the free energy, for the Edwards–Anderson order parameter q_{EA} and for the magnetization per site m. Ferromagnetic order has $m > 0$, while the spin glass phase has $m = 0$ and $q_{\mathrm{EA}} > 0$. The phase diagram of the spin glass according to SK is depicted in Fig. (9.7). SK realized from the outset that their solution leads to unphysical (i.e. negative) values of the entropy at low enough temperatures. They attributed the flaw to the interchange of the thermodynamic limit and the $n \to 0$ limit of the replica trick, and dismissed it as unimportant except at the lowest temperatures. It turns out, however, that the steepest descent method applied by SK does not correctly identify the maximum of the integrand, except for high temperatures. This was pointed out by de Almeida and Thouless [dAT78], who calculated the stability region of the SK method, and found that the spin glass phase entirely lay in the unstable region.

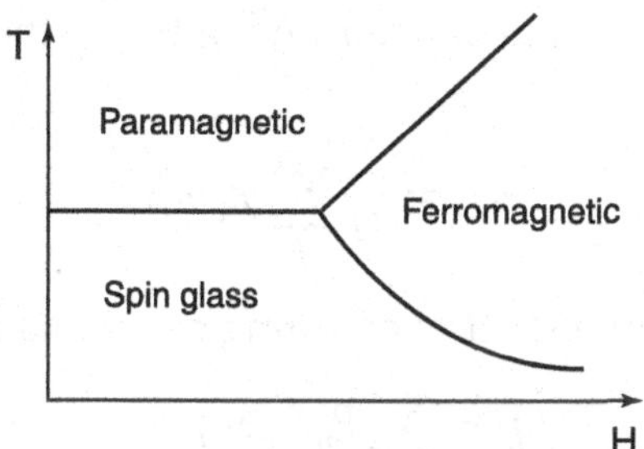

Figure 9.7. The phase diagram of the Sherrington and Kirkpatrick model. The relevance of this phase diagram for real spin glasses is hotly contested.

The Parisi solution

Parisi [Par79, Par80] pointed out that the replica symmetry had to be broken to avoid the instability of the SK model. In Parisi's approach, the order parameter of the spin glass becomes a piecewise continuous function $q(x)$ defined in the unit interval. Later [Par83], a physical interpretation for $q(x)$ was suggested in terms of the 'overlap' between so-called pure states of the spin glass.

Broken ergodicity refers to the inequality between time and ensemble averages encountered in systems thermally equilibrated below their critical temperature. A simple example of broken ergodicity is provided by the Ising ferromagnet. At $T = 0$, the ferromagnet has two ground states, say $+$ and $-$, which are connected by a global spin flip. At $T > 0$, and below the critical (Curie) temperature, the average magnetization in each of the corresponding so-called *pure states* is $\pm m$, where $m > 0$. The canonically averaged magnetization always vanishes, due to the symmetry of the Hamiltonian. In general, the canonical (Gibbs–Boltzmann) average of any observable quantity has the form:

$$\langle \cdots \rangle_{GB} = \sum_a P_a \langle \cdots \rangle_a, \tag{9.13}$$

where the average $\langle \cdots \rangle_a$ only includes the configurations belonging to pure state a.

Unlike the two pure states of the ferromagnet, the Parisi solution of the SK model features many pure states. Letting m_i^a be the magnetization of spin i in phase a, and defining the overlap between a and b as

$$q_{ab} = \frac{1}{N} \sum_i m_i^a m_i^b, \tag{9.14}$$

one defines, for a given set of coupling constants J,

$$P_J(q) = \sum_{a,b} P_a^J P_b^J \, \delta(q - q_{ab}), \tag{9.15}$$

where P^J_a is the probability of finding the system in its a'th pure state. Averaging over the disorder yields the function:

$$P(q) = \overline{\sum_{a,b} P^J_a P^J_b \, \delta(q - q_{ab})}. \tag{9.16}$$

By analogy with a ferromagnet, we imagine that a pure state could be singled out by *i)* applying a small field to a finite system, *ii)* letting the system size go to infinity and *iii)* letting the field go to zero, in this order.[4]

With reference to Eq. (9.4) we note that if the relaxing system is non-ergodic its trajectories remain within a single ergodic component or pure state. Hence, Eq. (9.3) for the Edwards–Anderson parameter can be rewritten as:

$$q_{\mathrm{EA}} = \overline{\sum_a P_a \, (m^a_i)^2}, \tag{9.17}$$

which is the average of the self-overlap of the magnetization, as well as the largest possible overlap between any two pure states. To express q_{EA} in terms of the Parisi function, we note that $P(q)$ is a probability density defined, and normalized, in the interval $[-1, 1]$. With a continuum of states both $P(q)$ and the corresponding distribution $x(q)$ are smooth functions of their arguments. Furthermore, x increases monotonically from zero to one, and has an inverse function – called $q(x)$ – which is just the fractile corresponding to a given x, e.g. $q(1/2)$ is the median of P, while $q(1)$ is the maximum possible overlap. By our previous argument $q(1)$ corresponds then to the Edwards–Anderson's order parameter. Unless the $P(q)$ is 'trivial', that is equal to a δ function, q_{EA} differs from the first moment $q^{(1)}$ of the Parisi function.

To estimate $P(q)$ [BY88], one considers two independently updated copies of the system (the replicas) with the *same* realization

[4]One difficulty of this gedanken process is that the nature of the local field is unclear for a random system.

of the bonds. After thermalizing the two systems using Monte Carlo updates,[5] one calculates the instantaneous overlap

$$Q(t) = \frac{1}{N}\sum_{i=1}^{N} S_i^1(t+t_0)S_i^2(t+t_0), \tag{9.18}$$

where, e.g., S_i^1 is the orientation of spin i in the first replica, and N is the number of spins. The Parisi function can then be estimated as:

$$P(q) = \overline{\frac{1}{t_{\max}}\sum_{t=1}^{t_{\max}} \delta(q - Q(t))}, \tag{9.19}$$

where the overbar denotes, as usual, a quenched average. This average is performed over many bond realizations, while the thermal average is performed over one long Monte Carlo trajectory.

Ultrametricity and hierarchies

The pure states of the SK model are 'ultrametrically organized'. To understand what this means, consider three different pure states, a, b and c, and define the distance between any of the two in terms of their overlap q_{ab} as:

$$d_{ab} = \frac{1}{2}(1 - q_{ab}). \tag{9.20}$$

The three states can then be depicted as the vertices of an isosceles triangle with a short base and two long sides. In other words, two of the distances are equal and the third is smaller, or else all distances are the same. Distances with this property are said to obey the 'strong ultrametric inequality'.

> **Exercise 9:** Ultrametricity and trees: Consider an upward rooted tree obtained by N successive bifurcations (multifurcations) branching out from the top node. The top node is level $N-1$ and the bottom nodes are at level 0. Assume that the states of a physical system are represented by the end nodes of the tree, and define the

[5] Unfortunately, this is only possible very close to the transition temperature.

distance between two zero'th level nodes as the level at which the branches emanating from the two nodes merge. Argue that this distance satisfies the strong ultrametric inequality.

Exercise 10: Not all trees are ultrametric structures: A widespread misunderstanding is that a tree whose nodes all correspond to physical states, e.g. the LS tree shown in Fig. (7.1), is an ultrametric structure. Why is this is not the case? (Hint: Find three states which do not fulfil the ultrametric inequality.)

Dynamical interpretation

The Parisi solution of the SK model is concerned with equilibrium properties and its pure states are by definition separated by infinite barriers. Hence they do not support dynamical processes. Nevertheless, the ultrametric structure of the SK model configuration space has inspired models where the pure states of the SK model are mapped into metastable states of a real system, and where their distance, which for the SK model is the Hamming distance, is mapped into an energy barrier. In this way, a model is created which has a hierarchically, and indeed ultrametrically, organized energy landscape, and which also lends itself to a dynamical interpretation [OS85, Sch85]. Using a vertical energy axis, the states of the model are all degenerate energy minima, sitting at the endpoints of the branches of an upward rooted tree [LOH$^+$91].

Exercise 11: In a regular ultrametric N level tree with branching ratio r and all states located at level 0, i.e. the bottom of the hierarchy. The ultrametric distance d_{ab} between states a and b, is the level at which the branches emanating from the two nodes join. We assume that moving from state a to b entails overcoming an energy barrier which grows with the ultrametric distance. Initially, the system is, with probability one, in one particular state, e.g. the leftmost node of the tree. Thermally activated transitions allow the probability distribution to spread over a growing number of states. Discuss the changes in average energy and entropy which this process entails. Is the result consistent with the fact that free energy decreases in a spontaneous process?

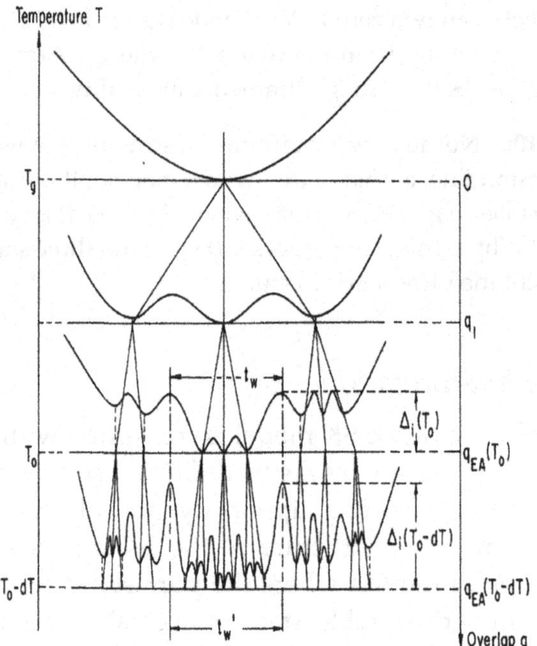

Figure 9.8. The picture, taken from Ref. [LOH+91], explains how, as the temperature is decreased, a hierarchical organization of metastable valleys can arise through a sequence of bifurcations.

Exercise 12: Does the above model contain the idea that ageing corresponds to an entrenchment into deeper and deeper valleys in the landscape? Why or why not?

Figure (9.8) is taken from Lederman *et al.* [LOH+91]. It is a graphical rendering of the hierarchical 'free energy' landscape[6] originally proposed by Ginzburg [Gin86] which, according to the above authors underpins hierarchical relaxation in spin glasses. Importantly, the structure of the landscape is parameterized by both temperature and time.

[6]In a free energy landscape, each point corresponds to a set of microscopic configurations. Each set supports a quasi-equilibrium state, and is usually characterized by the value of a suitable order parameter. The term is used liberally in the literature.

The first idea conveyed by the graph does not follow from, but rather supplements, the hierarchical organization borrowed from the SK model: As the temperature is lowered below T_g, the spin glass critical temperature, new valleys appear in a series of bi- or multi-furcations, i.e. new energy barriers appear, which are dynamically irrelevant at higher temperatures. The second idea is that the height of the barriers crossed at time scale t_w grows in the usual Arrhenius fashion. This hierarchical picture qualitatively accounts for the rejuvenation effects observed in spin glasses upon reheating a system previously subject to a small negative temperature step $-\Delta T$: Upon reheating the barrier structure present at the lower temperature and explored by ageing at that temperature simply disappears, leaving the system where it was before the cooling step.

The time spent at the lower temperature does not count at the higher temperature, as also shown experimentally in Ref. [LOH$^+$91] using TRM data. This result would not hold if the barriers delimiting the original valley could be overcome during the time spent at the lower temperature. The issue is circumvented in Ref. [LOH$^+$91] by assuming that the energy barriers separating different valleys grow much faster than linearly in ΔT. If a continuous distribution of barriers were present, the relevant barriers would scale linearly in both T and $\ln(t_w)$, a conclusion indeed supported by, e.g., the intermittency analysis given in Ref. [SJ05]. Also, the model does not seem to address the rejuvenation effect due to a small positive temperature step *not* preceded by a negative step of equal size:[7] This leads to randomization or 'chaos' [BM87] which in part or completely erases the memory of the current configuration. This is in contrast to the memory behaviour experimentally observed when reestablishing the status quo.

Exercise 12: The Local Density of States (LDOS) describes how the microscopic configurations belonging to a valley are distributed in energy. Figure (7.3) shows that the LDOS grows in a nearly exponential fashion with energy in an instance of the

[7] In spite of the author's claim to the opposite.

Travelling Salesman Problem. Similar behaviour has been found in the spin glass EA model [SS94, BS11a]. Neglecting any deviations from exponential growth, let us assume that in a range of energies $0 < E < \mathcal{L}$, $\mathcal{D}(E) = c\exp(E/T_0)$, where c is a constant. Show that the Boltzmann equilibrium distribution is 'bottom heavy', i.e. low energy states have high probability, for $T < T_0$. For $T > T_0$ high energy states are favoured. Show that a change of temperature across T_0 (in any direction) leads to redistribution of probabilities across the states, and hence to the destruction of any preexisting correlation patterns.

9.3.1. Theoretical Outlook

Our brief discussion of low temperature spin glass phenomena highlights features shared, as we shall see, by other complex systems and uncovers at the same time theoretical issues worth further consideration.

First a summary of the phenomenology: Spin glasses are clearly out of thermal equilibrium, and yet behave as if they were in equilibrium when observed over sufficiently short time scales. The boundary between equilibrium and non-equilibrium behaviour is the system age t_w, a quantity which affects all observations for a broad range of time scales. It has however been observed that the t_w dependence of the thermoremanent magnetization vanishes in the limit of very long observation time scales. Glossing for the moment over this last aspect, we note that the t_w dependence of linear response data already indicates a process of entrenchment in deeper and deeper valleys in the energy landscape. This hierarchical picture is supported by memory and rejuvenation effects. Its theoretical justification is connected, in the physics community, to a set of equilibrium properties of a mean field model, which are given a dynamical interpretation we find somewhat artificial. We argue that hierarchies of time scales, and hence, in a landscape context, of energy barriers are a very general paradigm for complex dynamics, and one which is unrelated to any particular spin glass model and only indirectly related to equilibrium properties. Real space properties are important in spin glasses and should not be left out of the picture: In real space, a partial thermal equilibrium establishes itself within small and slowly growing domains or clusters of contiguous spins. This forms the basis of a useful real space description of spin glass dynamics, which we have bypassed for space considerations. However, from a configuration space point of view, the presence of disjointed and statistically independent domains means that the energy landscape of a spin glass breaks down into an ensemble of energy landscapes, one corresponding to each domain. The dynamical evolution of such systems cannot hence be fully described in terms of the trajectory of a single point moving (stochastically) in a high dimensional energy landscape. This is the picture that we implicitly have used so far, and the picture must be amended in a certain respect: One needs to consider an ensemble of trajectories, each moving within a landscape, of its own. Even if different landscapes are statistically similar and can be modelled by identical structures, the initial conditions for the different trajectories can be different. Accordingly, physical quantities appear as averages over a distribution of initial conditions. This approach can explain all observed features of linear response functions, i.e. their scaling with system age and the 'end of ageing' effect [KBSR10, SK10].

Finally, a general consideration concerning how non-equilibrium ageing dynamics should be treated theoretically in complex systems. A popular route is to emphasize its similarity to equilibrium dynamics, i.e. try to extend the Fluctuation-Dissipation Theorem to non-equilibrium situation by using the concept of effective temperature [CKP97]. In some cases, this works well for large values of t and t_w, while in other cases it leads to unphysical value of the effective temperature [BBC03]. More importantly, the approach is intrinsically limited to systems such as spin glasses, where a linear response can be elicited in a laboratory setting, and where a state thermal equilibrium exists. This excludes, e.g., purely observational data and biological evolution processes. Furthermore, in our view the approach clouds an important issue in the relaxation dynamics of complex systems, namely the identification and statistical treatment of the rare non-equilibrium events visible in the tails of the fluctuation spectra discussed in this chapter. At a more abstract level, the issue is how metastable attractors are dynamically selected during an ageing process.

10

Magnetic Relaxation in Superconductors

10.1. INTRODUCTION

This section deals with the motion of magnetic flux lines inside superconductors and highlights how certain aspects of their dynamics are akin to other examples of the slow non-stationary relaxation discussed in this book.

A magnetic field inside type II superconductors behaves in many respects like a material with its own elastic and plastic properties. The field forms flux tubes, or flux lines, that penetrate the superconducting material and interact with each other and with inhomogeneities in the material. In traditional superconductors, which operate at temperatures below 10 K, thermal fluctuations have little importance for the flux line structure. Since the discovery of the high temperature superconductors in 1986 [BM86] (for a general overview see [Bur92]) the effects of thermal excitations on the structure, relaxation and dynamics of flux lines have been explored in a wealth of experiments. How superconductivity works is described in many good books such as [Tin04, Bur92]. Here we focus on the aspects of flux matter relevant for our discussion. We first briefly summarize the key aspects of how the magnetic field behaves inside type II superconductors and then describe a simple model which captures a surprising aspect of the thermally activated dynamics of the flux lines as they move in and out of the samples. Namely that the magnetic relaxation rate depends only weakly on the temperature, see, e.g., [CMM+90, KKM+90].

10.2. MAGNETIC FIELDS AND TYPE II SUPERCONDUCTORS

Type II superconductors remain superconducting even after the applied magnetic field H begins to enter the sample at a field strength $H > H_{c_1}$. When H exceeds a certain value $H_{c2} > H_{c_1}$ the material becomes a normal conductor. For H between H_{c_1} and H_{c_2}, the external magnetic field penetrates into the bulk of the superconductor in a spatially non-uniform pattern in which the field is concentrated along vortex or flux lines created by circulating currents, or vortices of superconducting electrons.

Parallel flux lines repel each other with an energy that depends on their distance, and furthermore they interact with inhomogeneities in the material: At the centre of the flux lines the material is in its normal phase, and any inhomogeneity which favours the normal phase will therefore tend to attract the flux line core. This leads to the following situation: When the external magnetic field is raised to values above H_{c_1} magnetic flux lines enter the superconducting material from the surface. Since flux lines repel each other, they push each other towards the bulk of the material. If the material is homogenous, i.e. without impurities, a uniform density of flux lines is rapidly established with a local density determined by the magnetic pressure exerted by the external magnetic field. The presence of inhomogeneities tends to trap the flux lines and thereby prevents the rapid establishment of a uniform flux density. A mechanical force balance may be established between the pinning force exerted by an inhomogeneity, also called a pinning centre, and the force produced by the gradient in the surrounding flux line density. The pinning centre corresponds to a local potential energy well for the flux line. At non-zero temperature thermal fluctuations can release a flux line from this well and allow it to move towards regions of lower flux density, as a result kicking other flux lines out of their traps. In this way, flux avalanches can be created which lower the gradient in the density of flux lines as these move into the bulk region of the material. As a consequence, the difference between the internal magnetic field and the external applied field decreases. This process is called magnetic relaxation since the magnetic state of the material *relaxes* towards its thermodynamic equilibrium state.

The rate at which the magnetic flux lines are able to overcome the hindrances to their motion due to the pinning centres should depend strongly on the temperature. If we imagine that individual flux lines are trapped into local energy wells and that they exit these traps as a result of thermal fluctuations, the probability per time unit for a flux line to exit by overcoming the corresponding energy barrier ΔE is proportional to $\exp(-\Delta E/(k_b T))$, and has a strong exponential dependence on the temperature T. The experimental observation that the magnetic relaxation rate, or creep rate, only varies weakly in a broad range of temperature in high temperature superconductors, see [CMM$^+$90, KKM$^+$90], therefore came as a great surprise. Several explanations for this unexpected behaviour have been put forward, including collective pinning effects and quantum tunnelling of flux bundles. Here we discuss how the phenomenon is simply explained in terms of record dynamics. To this end, we introduce in the next section a model of flux line dynamics which only includes the bare essentials of the lines mutual interactions and interactions with thermal noise.

10.3. RESTRICTED OCCUPANCY MODEL

The Restricted Occupancy Model (ROM) provides a simplified coarse-grained description of the electromagnetic dynamics of type II superconductors. It is related to earlier lattice models of flux motion at zero temperature, see [Jen90, BP98, BPR99] but differs from them in several important respects. The ROM is defined either on a two-dimensional lattice or on a three-dimensional stack of layers. It takes into account the existence of an upper critical magnetic field by having a maximum number of flux lines per lattice site, and its Monte Carlo dynamics is defined at all temperatures. The model was first studied in two dimensions, where its thermal dynamics is able to describe a range of important aspects of vortex dynamics at non-zero temperature [JNP+00, NJ00, NJ01b, NJ01a, JN01]. Here we describe how its three-dimensional version deals with the long-standing puzzle of the observed weak temperature dependence of the creep rate.

The interaction between flux lines decays over a length scale given by the London penetration depth λ. This length scale is a material specific quantity and is often very long, e.g. of the order several thousand of Ångström. The distance between flux lines is controlled by the magnetic induction since each flux line carries one quantum of flux. The average separation of flux lines at a magnetic field strength of 1 Tesla is about 450 Ångström, hence a flux line easily interacts with many other flux lines.

High temperature superconductors are layered materials. When an external magnetic field is applied perpendicularly to the layers one may envisage the flux lines as passing through a stack of vortices or 'pancakes', one for each superconducting layer. The ROM model does not keep track of the detailed motion of the stacks of pancakes. Instead, a discrete grid is introduced for each superconducting layer and the number of flux lines n_i transversing the i'th plaquette of the grid is used as a dynamical variable, see Fig. (10.1). The

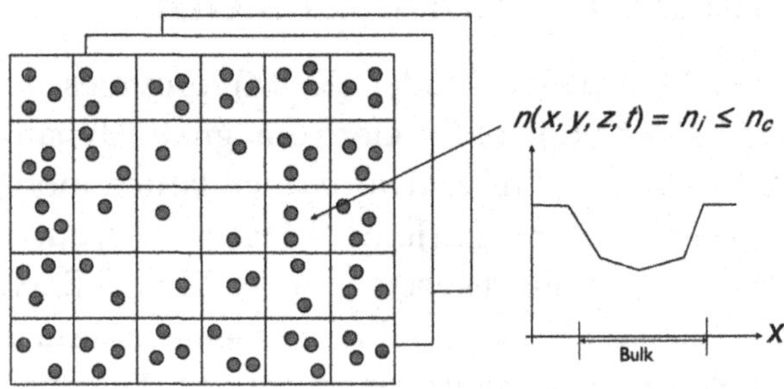

Figure 10.1. Sketch of the layered ROM model and the profile of the magnetic field across the sample.

linear size of the plaquette equals the interaction length scale λ and the only interactions included are those between flux lines in the same or nearest neighbour plaquette. This means that the motion of the flux lines is represented by a time variation in the numbers n_i. The interaction between the flux lines is modelled by an energy function, the Hamiltonian, which is expressed in terms of the configuration $\{n_i\}$. Namely,

$$H = \sum_{ij} A_{ij} n_i n_j - \sum_{i} A_{ii} n_i + \sum_{i} A_i^p n_i + \sum_{\langle ij \rangle_z} A_2 \left(n_i - n_j\right)^2 . \quad (10.1)$$

The first term represents the repulsion energy due to vortex-vortex interaction within a layer, and the second the vortex self energy. As already mentioned, interactions beyond nearest neighbours are neglected, since the potential that mediates this interaction decays exponentially at distances longer than our coarse-graining length λ. We set $A_0 \stackrel{\text{def}}{=} A_{ii} = 1$, $A_1 \stackrel{\text{def}}{=} A_{ij}$ if i and j are nearest neighbours on the same layer, and $A_{ij} = 0$ otherwise. The third term represents the interaction of the vortex pancakes with the pinning centres. A^p is a random potential and we assume for simplicity that A^p has the following distribution $P(A^p) = (1-p)\delta(A^p) - p\delta(A^p - A_0^p)$. The pinning strength $|A_0^p|$ represents the total action of the pinning centres located on a site. In the simulations described here $A_0^p = -0.3$. Finally the last term in Eq. (10.1) describes the interactions between the vortex sections in different layers. This term is a nearest neighbour quadratic interaction along the z-axis, so that the number of vortices in neighbouring cells along the z-direction tends to be the same.

The parameters of the model are defined in units of A_0. The time is measured in units of Monte Carlo (MC) sweeps. Further details concerning the relationship between the model parameters and material parameters are discussed in [NJ00, JN01]. We assume the external magnetic field to be applied perpendicularly to the planes and for this reason we only consider motion of the vortex pancakes within the planes and use periodic boundary conditions in the z-direction. Each individual Monte Carlo update consists in selecting a vortex pancake at random and moving it to a randomly selected neighbour position. As always in Metropolis importance sampling the movement of the vortex is automatically accepted if the energy of the system decreases or remains unchanged. If the energy of the system increases, the movement is accepted with probability $\exp(-\Delta E/T)$.

The magnetic pressure exerted by the external field is controlled by a time dependent field $H_{\text{edge}}(t)$ which only acts on the sites along the edges of each plane. Initially this external field is rapidly increased to a desired value, (in the simulation discussed below, $H_{\text{edge}} = 10$ pancakes per edge site). During a MC sweep pancakes may move between edge and bulk sites, but after each sweep the density on the edge sites is again brought back to the relevant value $H_{\text{edge}}(t)$. For further details consult [JNP$^+$00, NJ00, NJ01b, NJ01a, JN01]. After a fast initial ramping, the external field is kept constant. We study how the vortices move into the sample and find that in a very broad range of temperatures, the pancakes from the edges enter the bulk through intermittent bursts. Below we discuss the statistics of the bursts' sizes and of the times at which bursts of pancakes enter the bulk.

10.4. INTERMITTENT MAGNETIC RELAXATION

We now turn to the dependence on time and temperature of the total number of pancakes in the sample. In Fig. (10.2) the number of pancakes in the bulk is seen to increase in a stepwise fashion. One notices that the steps occur at times that appear to be uniformly distributed along the logarithmic time axis. This is the first indication that the time series consisting of the time instances at which the vortex pancakes enter the system may follow record dynamics. In Refs. [PAS04, OJNS05] it was indeed demonstrated that the statistics of these entrance times is Poissonian with an average depending on the logarithm of time. This being the case, the steps, which correspond to a cluster of flux lines entering the bulk, will be called *quakes* in accordance with the notation in the rest of the book.

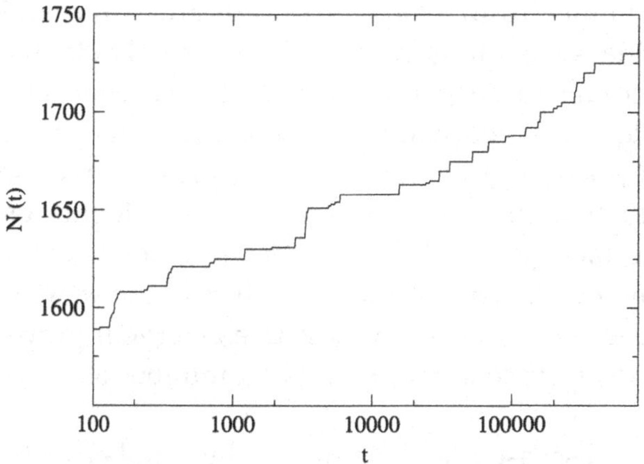

Figure 10.2. The variation of the total number of vortices, $N(t)$, on the system as function of time t, for a single realization of the pinning potential and the thermal noise in a $8 \times 8 \times 8$ lattice for $T = 0.1$. Notice that the steps seem to be uniformly distributed along the *logarithmic* time axis as expected for a Poisson process in logarithmic time. The figure is taken from Ref. [OJNS05].

The interpretation in terms of record dynamics implies [SL93, SD03] that the probability that exactly q quakes occur during the time interval $[t_w, t_w + t]$ is given by

$$p(q) = \frac{\langle q \rangle^q}{q!} \exp(-\langle q \rangle), \tag{10.2}$$

where the average $\langle q \rangle$ has a linear dependence on logarithmic time

$$\langle q \rangle = \alpha \log((t + t_w)/t_w). \tag{10.3}$$

For consistency with the notation of Ref. [OJNS05], the symbol t stands in this chapter for the time elapsed after t_w, rather than for the time elapsed since the beginning of the relaxation process, which is the convention preferred in this book.

The value of the pre-factor α in Eq. (10.3) is estimated to be $\alpha = 22.6$ from the logarithmic rate of the quake events observed in the simulations. Its independence of the temperature T lies at the root of the temperature independence of the creep rate. Mathematically it stems from the fact that record statistics is independent of the noise distribution from which the records are extracted. Physically, reducing the temperature rescales the energy landscape in which the flux lines move, i.e. part of the landscape becomes dynamically inaccessible. However, within the part which remains accessible, new traps previously dynamically irrelevant become now important, i.e. fluctuations which are reversible at a higher temperature become de facto irreversible and therefore count as quakes at a lower temperature. The self-similar nature of the energy landscape in which flux lines move is a collective emergent property, which requires the presence in the system of a number of interacting flux lines.

As later discussed in more detail, the number of flux lines in the system is a distributed quantity, which depends on both the number of quakes the system experiences and on the number of flux lines entering during each quake. The latter quantity is T independent. On average, and banning initial transients, the number present in the system and hence the vortex line density, is then expected to increase in proportion to the logarithm of time and in a temperature independent fashion. Figure (10.3) shows that the vortex number for a range of temperatures is a piecewise linear

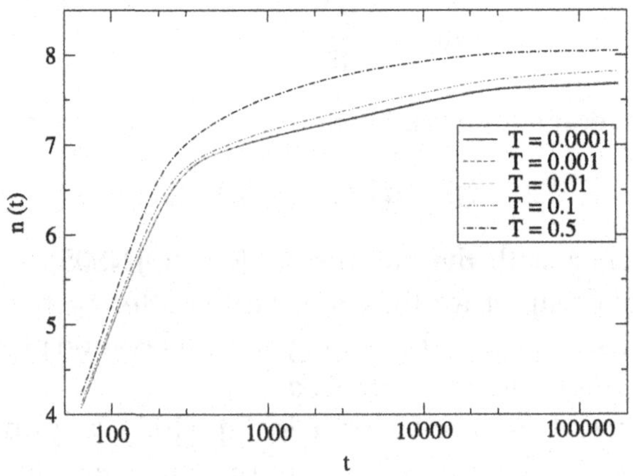

Figure 10.3. The density of flux lines $n(t)$ is plotted versus time for various temperatures. For $T \leq 0.1$ the vortex number is a piecewise linear function of $\log(t)$. For $T = 0.5$ the system relaxation becomes faster. Figure taken from Ref. [OJNS05].

function of $\log(t)$, with three different regimes separated at times $t_1 \approx 300$ and $t_2 \approx 3 \times 10^4$. The first regime has only a few flux lines inside the system, and describes the initial transient already alluded to. In the second regime, $t_1 < t < t_2$, vortex interactions become essential. We focus on this regime, which is closest to the predictions of record dynamics, and take, in the following, $t_w = t_1$. The late time regime $t > t_2$ shows signs of saturation, but is very difficult to resolve appropriately in simulations and probably equally difficult to study experimentally.

We now consider the properties of the total number of vortices $N(t_w + t)$ present in the system at time $t_w + t$. Let t_k denote the time of occurrence of quake number k and let v_k denote the actual number of vortices entering during this quake. We then have

$$N(t + t_w) = N(t_w) + \sum_{t_w < t_k < t_w + t} v_k, \tag{10.4}$$

where the sum is over all quakes that occurred during the observation time $[t_w, t_w + t]$.

We assume that the number of pancakes v which enter during a given quake (see insert Fig. (10.4)) has an exponential distribution $p(v) = \exp(-v/\bar{v})/\bar{v}$, and further assume that consecutive quakes are statistically independent. The number of vortices entering through q quakes is then a sum of independent and exponentially distributed stochastic variables, and is hence Gamma distributed. To obtain the PDF of the total number of vortices entering during $[t_w, t_w + t]$, this Gamma distribution must be averaged over the probability, Eq. (10.2), that precisely q quakes occur during the time interval of interest. This leads to the following expression for the PDF of total number of vortices $\Delta N = N(t + t_w) - N(t_w)$ entering during the time interval $[t_w, t_w + t]$:

$$p_{\Delta N}(x, t) = e^{-\frac{x}{\bar{v}} - \langle q \rangle} \sqrt{\frac{\langle q \rangle}{\bar{v} x}} I_1\left(2\sqrt{\frac{\langle q \rangle x}{\bar{v}}}\right), \tag{10.5}$$

where I_1 denotes the modified Bessel function of order 1. This theoretical prediction, is compared in Fig. (10.3) with our simulation results.

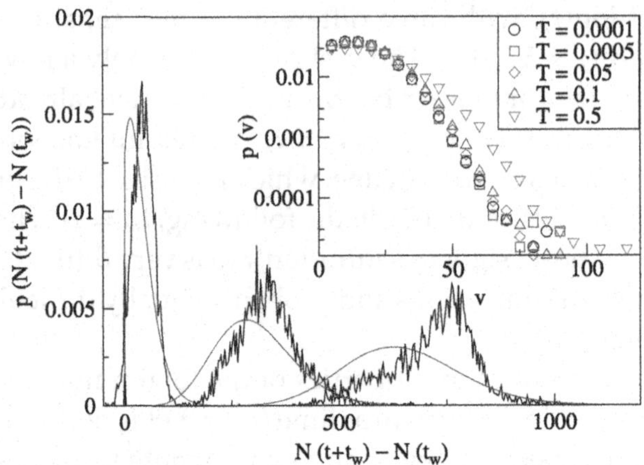

Figure 10.4. The main panel contains the temporal evolution of the probability density function, $P(N(t + t_w) - N(t_w))$, of the number of vortices entering for $t_w = 1000$ and three different observation times $t = 188, 2791, 8371$ given by the slightly jagged curves. The smooth curves are a fit to the theoretical expression (see Eq. (10.5)). The system is $16 \times 16 \times 8$ and $T = 0.0001$. The insert shows the quake size distribution for various temperatures for the time interval between $t = 1000$ and $t = 10000$. For $T \leq 0.1$ the distribution has an approximately exponential tail. For $T = 0.5$ the shape gets closer to a Gaussian distribution. Figure from [OJNS05].

We can determine the average $\bar{v}$ in two ways. Either directly from the simulated distributions in the insert of Fig. (10.4) or from fitting Eq. (10.5) to the simulated data shown in Fig. (10.3). In both cases we find $\bar{v} = 16$. We also find that $\bar{v}$ is essentially temperature independent for temperatures below $T \approx 0.1$. This is also seen from the insert in Fig. (10.4). Importantly, the MC dynamics does overcome plenty of positive energy barriers, $\Delta E > 0$, through thermal activation for temperatures in the range $0.01 < T < 0.1$. As the temperature is lowered fewer MC updates correspond to $\Delta E > 0$ and for the lowest temperatures MC steps mainly involve $\Delta E \leq 0$ only [NJ01a]. Nevertheless, the record dynamics remains essentially temperature independent for $T < 0.1$.

Figure (10.4) shows that v_k has a temperature independent distribution for $T < 0.5$. Since the average number of quakes increases linearly with logarithmic time, see Eq. 10.3, record dynamics predicts the following temperature independent temporal evolution of the average number of vortices

$$\Delta N \equiv \langle N(t+t_w)\rangle - \langle N(t_w)\rangle = \alpha\bar{v}\log(1+t/t_w). \qquad (10.6)$$

I.e. for the considered time regime $t/t_w \gg 1$ a temperature independent logarithmic creep rate given by

$$d\Delta N/d\log(t) = \alpha\bar{v}/(1+t_w/t) \approx \alpha\bar{v}. \qquad (10.7)$$

We extract the rate of the quakes in the simulations from temporal signals like the one exhibited in Fig. (10.2) and find the rate to be independent of temperature in the broad temperature interval $10^{-4} < T < 2\times10^{-2}$. We are now able to compare this behaviour of the ROM model with experimentally measured magnetic creep rate, see Fig. (10.5) which shows the near temperature independence of

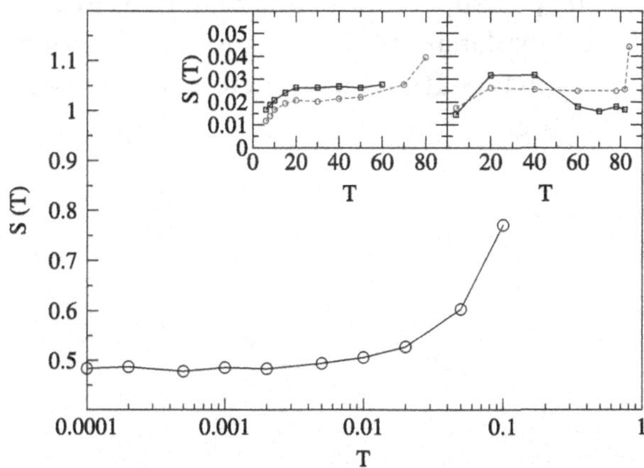

Figure 10.5. Main panel: Numerical results for the creep rate versus T for the time interval between $t = 1000$ and $t = 10000$. In the low temperature region the creep rate is constant within our numerical precision for about two orders of magnitude; we observe a non-zero creep rate in the $T \to 0$ limit. Insets: Experimental results for the creep rate versus T. The right inset shows data from Keller *et al.* [KKM+90] for melt processed YBCO crystals with the magnetic field applied along the *c* axis (squares) and *ab* plane (circles). The left inset shows data from [CMM+90] for unirradiated (squares) and 3 MeV proton irradiated (circles) YBCO flux grown crystals with a 1 T magnetic Field applied parallel to the *c*-axis. Figure from [OJNS05].

the actual rate with which $N(t)$ changes. We obtain this rate from the data shown in Fig. (10.3) in the time region $t_1 < t < t_2$ and plot the normalized creep rate $S = d\log[M(t)]/d\log(t)$ in order to compare consistently with experiments, we have used $M(t) = |H(t) - H_{\text{edge}}|$.

In the ROM model the movement of flux lines into the bulk is thermally activated, i.e. controlled by thermal energy fluctuations, but the observed magnetic creep rate has nevertheless a very weak temperature dependence. The vortices enter the bulk in an intermittent fashion through quakes, irreversible events with a log-Poisson statistics which is independent of the noise and hence independent of temperature. This allows us to conclude that the magnetic relaxation process in the ROM is controlled by record-sized thermal noise fluctuations which trigger irreversible movements in a hierarchical energy landscape. The same could be true for magnetic flux creep in superconducting materials. Record dynamics thus provides an explanation of the temperature independence of the flux creep rate in type II superconductors which does not involve quantum tunnelling.

11

Ageing of Colloids

11.1. INTRODUCTION

In a colloid, small particles of linear scale ranging from tens to thousands of nanometres are dispersed in a fluid. Colloidal particles can interact with one another and with the surrounding fluid in a variety of ways, e.g. Laponite suspensions consist of charged platelets which interact electrostatically, while interactions in hard sphere colloids are repulsive and very short-ranged. In addition, gravity affects all components in a colloidal suspension, creating, over time, density gradients and sedimentation.

The density of the colloidal particles is generally expressed in terms of the volume fraction ϕ, i.e. the fraction of the sample volume occupied by the particles. As ϕ is increased towards a critical value ϕ_c the sample viscosity increases. Concomitantly, the time scales characterizing various physical relaxation processes all increase, e.g. the time scale associated with particle displacements of a certain length increases, leading to a decrease of the diffusion coefficient. This behaviour is in many ways similar to that of glass forming liquids cooled towards the glass temperature T_g. Colloids are therefore used as model systems for the glass transition, and a large body of theoretical work, see, e.g., [BF99], is dedicated to the calculation of their critical density. It should be noted that while the viscosity change seen in colloids approaching ϕ_c is large, it remains several orders of magnitude smaller than the corresponding change observed in glass forming liquids approaching T_g. Possibly, the difference arises because colloidal density is experimentally harder to control and fine-tune than temperature in a liquid glass former.

Since ageing dynamics is discussed in this book in different contexts, the terminology needs some clarification at this point. Many colloidal studies describe the systematic age dependence of the relaxation time scales observed at densities below ϕ_c. Such effects can generally be expected in any complex system with a wide spectrum of relaxation time scales, as the system age acts then as a low frequency cut-off for the dynamically relevant modes, i.e. modes which have not yet decayed at t_w. The spin glass magnetic susceptibility data of Svedlindh *et al.* [SGN+87], redrawn in our Fig. (9.1), show the temperature dependence of the susceptibility for a wide range of frequencies. Above as well as below the spin glass transition temperature, which in this case is $T_c = 22.6$ K, the susceptibility depends on the frequency ω of the magnetic field. For $T > T_c$, data taken in a low frequency range $\omega \leq \omega_{\text{min}}(T)$ collapse, showing that the system reaches equilibrium on a time scale $t(T) = 1/\omega_{\text{min}}(T)$. Furthermore, $t(T)$ diverges for $T \downarrow T_c$ and becomes undetectable at sufficiently high temperatures. Below T_c, there is no sign of equilibration, i.e. the susceptibility curves taken at different frequencies all remain well separated at all temperatures. In the spin glass literature the main focus is on the ageing behaviour below T_c and the term 'ageing dynamics' refers to the characteristic low temperature non-stationary regime, which is observed in colloids for $\phi > \phi_c$. The term 'ageing' is used more broadly in colloids, i.e. whenever age dependent effects are observed. While in this book we generally stick to the more restrictive use of the term, we strive in this chapter to clarify how the term is used as need arises.

In the broad sense of the word, ageing effects are measured in colloids by static and dynamic light scattering data, viscoelastic properties, electric noise fluctuations and by collecting time traces of particle motion by means of confocal microscopy. The following section focusses on hard sphere colloids and relies on results obtained by dynamic light scattering and confocal microscopy. It comprises two subsections, respectively dealing with properties below and above the critical density. The discussion below leans heavily on a (small) number of original references. These should definitely be consulted for further experimental details and for a more in-depth discussion than space allows here. For a general introduction to colloidal dynamics we refer the interested reader to a recent review by Hunter and Weeks [HW11].

11.2. HARD SPHERE COLLOIDS

Hard-sphere colloids are particularly interesting from our point of view, as their very short-ranged and repulsive interactions do not allow the formation of nested valleys organized in a hierarchical energy landscape of the sort controlling the dynamics of spin glasses, i.e. all available configurations of a colloid have essentially the same energy.

In this sense, dense hard sphere colloids find themselves at one extreme of a range of different ageing systems, with spin glasses, whose dynamics is thermally activated, being at the opposite end. From a simulation view point, hard sphere colloids are simple to define and simpler to deal with than systems featuring interactions with a longer range. It should be stressed, however, that 'hard sphere' behaviour in real colloidal systems is not easily achieved. Experimental samples have short-ranged attractive interactions which would make the particles clump together and even crystallize, and gravitation induces sedimentation which affect measurements taken over very long times. All these problems must be avoided or at least mitigated in experiments. Leaving technical challenges aside, we consider two basic questions. The first regards how the relevant relaxation time scales grow as the density approaches its critical value. As we discuss in the following, relaxation time scales characterizing the approach to equilibrium do possess a clear age dependence, even though the system is not ageing in the (restrictive) sense of the term used in this book. The second question regards the dynamical properties beyond the critical density, in the truly non-ergodic regime where equilibrium is completely out of reach within experimentally accessible time scales.

System age t_w can be defined precisely in a computer simulation as the time elapsed since an instantaneous system change, e.g. in a computer code one can instantaneously change an interaction parameter and create a colloidal suspension from a liquid. The far more laborious preparation procedure for experimental systems aims at producing a spatially homogeneous sample with no air bubbles and includes several hours of vortexing, centrifuging and tumbling. The system age is generally measured from the end of the tumbling stage.

11.2.1. The Critical Density

Hard sphere colloids have been extensively investigated by dynamic light scattering techniques in the interesting work of El Masri *et al.* [EMBP+09], a work which also includes relevant simulation results. In our discussion below, the experimental results obtained using dynamic light scattering stem from this reference, including our Figs (11.1) and (11.3). The important Intermediate Scattering

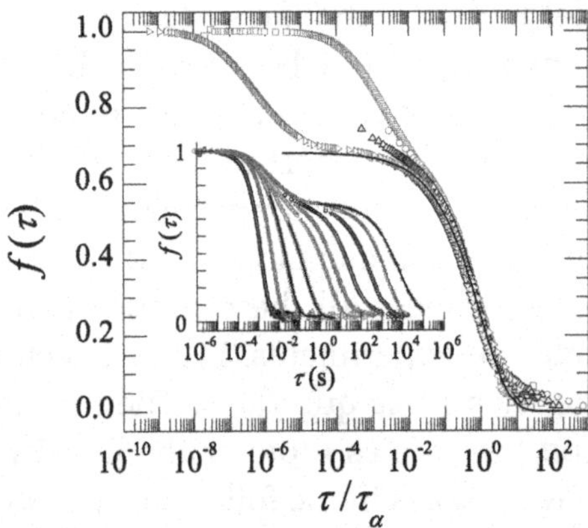

Figure 11.1. Insert: The Intermediate Scattering Function (ISF) is plotted vs. the lag time τ for different volume fractions. From left to right: $\phi = 0.0480, 0.3096, 0.4967, 0.5555, 0.5772, 0.5818, 0.5852, 0.5916, 0.5957$ and 0.5970. Main: All curves for $\phi \geq 0.4967$ are collapsed using the scaled variable $\tau/\tau_\alpha(\phi)$, where $\tau_\alpha(\phi)$ is approximately given by Eq. (11.2). The full line shows the stretched exponential $f_{\text{fit}}(\tau) = 0.60 \exp(-(\tau/\tau_\alpha)^{0.558})$. Figure taken from Ref. [EMBP+09].

Function (ISF), $f(t_w, t_w + \tau)$ is a measure of correlation between the fluctuating positions of the particles at times t_w and $t_w + \tau$. Note that τ is a time difference or lag between two observations, rather than the time counted from the beginning of the process. In the low density interval where the colloidal systems can reach equilibrium, the ISF decay is empirically well described by the stretched exponential function

$$f(t_w, t_w + \tau) = A(t_w) \exp\left[- \left(\frac{\tau}{\tau_\alpha(\phi, t_w)} \right)^{p(t_w)} \right] + B. \tag{11.1}$$

The formula describes a slow structural relaxation process, so-called α relaxation, p is the stretching exponent, and A and B are constants. Experimentally, the relaxation of colloidal hard sphere systems is characterized by the time scale $\tau_\alpha(\phi, t_w)$, a time scale which depends on the volume fraction as well as on the system age. For any fixed density below ϕ_c, τ_α increases monotonically with t_w, but only up to a constant or plateau value. The latter increases monotonically as a function of the volume fraction ϕ. The beginning of the plateau coincides with the equilibration time scale, a time scale increasing with density over approximately two orders of magnitudes, i.e. from $\approx 10^3$ for $\phi = 0.5876$ to $\approx 5 \times 10^4$ seconds for $\phi = 0.5970$. For $\phi = 0.5990$, a plateau cannot be clearly discerned from the data, whence one concludes that, empirically, the critical volume fraction is $\phi_c \approx 0.6$. Below ϕ_c, equilibrium can be reached, and the t_w dependence of τ_α can thus be removed by waiting long enough. For the corresponding equilibrium relaxation time scale, we simply write $\tau_\alpha(\phi)$. The same applies to other quantities which enter Eq. (11.1): They all lose their age dependence in the equilibrium fluctuation regime, i.e. the equilibrium ISF can be written as $f(\tau)$. In this time-translation invariant regime, Fig. (11.1) fully confirms Eq. (11.1): Scaling the abscissa with $\tau_\alpha(\phi)$ collapses most of the data shown in the insert into a single curve. El Masri *et al.* show how the time scale $\tau_\alpha(\phi)$ used in the scaling increases with ϕ just below $\phi \approx 0.6$ as

$$\tau_\alpha(\phi) \propto (\phi_c - \phi)^\gamma, \tag{11.2}$$

where $\gamma \approx 2.5$. Confocal microscopy produces images with a narrow field depth, e.g. images where only a thin horizontal slice of the sample is in focus. Scanning through focal planes located at different heights and using specialized software, the colloidal particles can all be tagged and their coordinates identified. Iterating this (laborious) procedure over time yields the time dependent spatial trajectories of the particles, whence different statistical measures characterizing correlation structure and particle motion can be estimated. Since relaxation time scales are finite and increasing with density below ϕ_c, a diffusive behaviour can be expected, with a diffusion coefficient decreasing as $\phi \to \phi_c$ [EMBP+09]. Our Fig. (11.2) is taken from Ref. [BS11b], and depicts on linear scales the MSD extracted from the original experimental data of Courtland and Weeks [CW03]. The MSD is given in square micrometres (μ^2) and the time in hours. Different curves correspond to different values of the volume fraction, from top to bottom $\phi = 0.46, 0.52, 0.53$ and 0.56. As anticipated, the diffusion coefficient, which is the slope of the different lines, decreases for increasing ϕ.

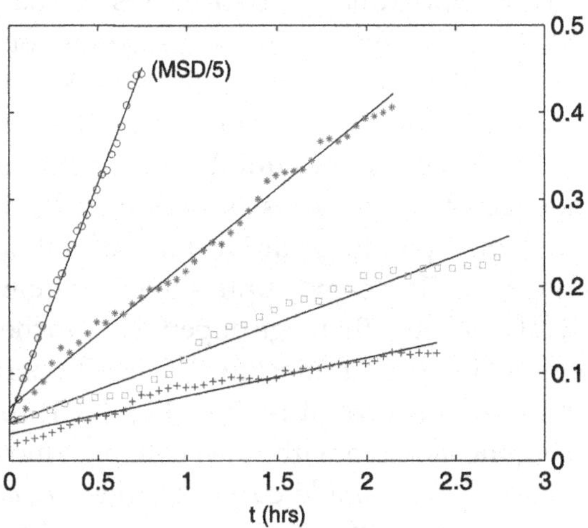

Figure 11.2. The Mean Square Displacement (MSD) over time t averaged over all particles. From top to bottom $\phi = 0.46, 0.52, 0.53$ and 0.56. The figure is taken from Ref. [BS11b] and is based on the experimental data by Courtland and Weeks [CW03].

11.2.2. Non-ergodic Ageing Behaviour

Above the critical density, the non-ergodic ageing dynamics is characterized by a decelerating tempo, i.e. decorrelation and particle movements occur at a slower rate the larger t_W is. Beyond that, the behaviour is less well understood than say, comparable behaviour in spin glasses, since no general agreement seems to exist on how correlation functions and particle displacement scale with system age. The decelerating nature of ageing dynamics is clear in Fig. (11.3). At any fixed value of the lag time τ, the decay of the ISF becomes gradually smaller as the age t increases from 500 s to approx 48 hrs, an increase of more than two orders of magnitude. At the lowest value of t_W the decay is logarithmic in most of the range. Note that, according to a well-established tradition, the abscissa is the lag time and not the time counted from the beginning of the ageing process. This choice is natural when the dynamics possesses time translational invariance, but lacks a theoretical underpinning in the ageing regime.

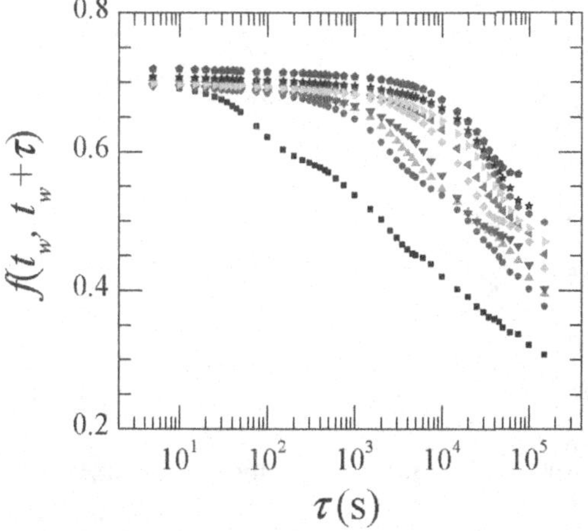

Figure 11.3. Insert: The Intermediate Scattering Function (ISF) in a dense colloid (volume fraction $\phi = 0.5990$) is plotted on linear-log scales vs. the lag time τ. Different curves correspond to different ages, from bottom to top $t_W = 10^3(0.5, 5, 7, 10, 20, 30, 50, 70, 100, 170)$ s. Figure taken from Ref. [EMBP+09].

Whether ageing affects the static radial correlation function or any other static measures connected to the distribution of colloidal particles in space is a natural question to ask, considering that in, e.g., spin glasses the real space signature of ageing is precisely the growth of thermally correlated domains [Rie93]. Cianci *et al.* [CCW06] investigated the issue for dense colloids and answer the question in the negative. Using colloidal particle positions obtained by confocal microscopy, they study the static structure of the sample by determining how particles are packed in irregular tetrahedra. These tetrahedra have attributes, e.g. mobility, volume, irregularity, whose distribution through the sample they studied as a function of age. The mobility clearly decreases with system age, but the geometrical properties of the tetrahedra do not reflect the age of the colloid. Returning to the statistics of particle motion in dense colloids, Fig. (11.4) is taken from Ref. [BS11b], where the original experimental data by Courtland and Weeks [CW03] were reanalysed. It shows how the MSD in a dense colloid appears simple when plotted a function of the time variable $t = t_w + \tau$ scaled by the waiting time t_w.

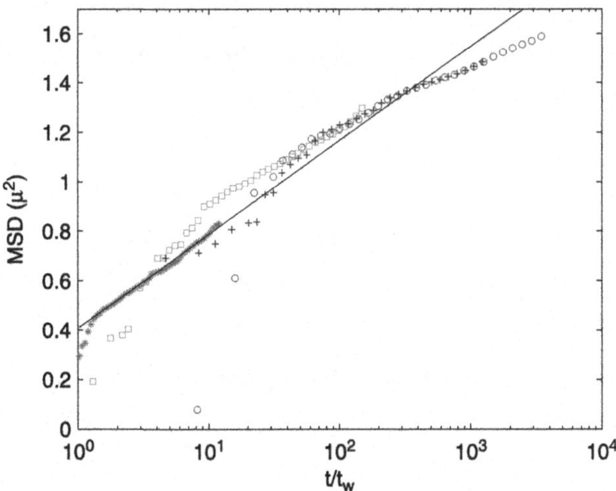

Figure 11.4. The MSD (in μm^2) dependence on the ratio t/t_w is nearly logarithmic for volume fractions above the critical density. Here, the volume fraction is $\phi \approx 0.62$. Plusses, squares, circles, and stars correspond to $t_w = 1, 10, 20$ and 150 s, respectively. Figure taken from Ref. [BS11b].

Note that t_W and t are measured from the same initial point, namely the time at which the stirring of the sample ceases. The values of t_W used in the analysis are however somewhat uncertain. On the basis of these results, Ref. [BS11b] (unconventionally) argues that particle motion in dense colloids can be seen as a diffusion process in logarithmic time, i.e. that dense colloids are yet another system where the change of variable from time to its logarithm renders the dynamics translationally invariant. In this perspective, colloidal dynamics appears even simpler than spin glass dynamics: On a logarithmic time scale, the position of each particle is merely the sum of independent increments. What triggers these increments and how their statistical independence arises is far from obvious from the MSD data presented above. A possible connection with record dynamics is explored through the model [BS11b] briefly described in the next section.

11.3. A HEURISTIC MODEL

While ageing is naturally seen as a relaxation process, identifying which physical properties actually relax in a hard sphere colloid is, as mentioned, surprisingly difficult [CCW06]. The model discussed below was developed in Ref. [BS11b] and assumes that what ages is a pattern of strong kinematic constraints binding neighbouring particles together into clusters whose lifetime increases with their size. As long as a cluster survives, its particles only move in a correlated fashion, and the cluster's centre of mass merely fluctuates in time around a fixed spatial position. Eventually, random forces from the surroundings destroy the cluster, allowing the particles to take a random step in a random direction, thereby joining adjacent clusters. This description implies a very high degree of spatial heterogeneity: While a minority of very mobile particles is involved in cluster rearrangements localized in different areas of the system, the vast majority stays put.

Along these lines, the one-dimensional computer model considered in [BS11b] deals with a circular array of points on which clusters can form. The update dynamics is a Markov chain of length L

where a randomly selected cluster of size h either survives intact or is destroyed with probability $0 \le P(h) \le 1$. In the model, larger clusters are more stable than smaller ones, i.e. $P(h)$ decreases with h. The stability of clusters depends on details of the short-range interactions between the particles to which we do not have access. We note however that if, in a real system, a cluster has had sufficient time to grow large, the pattern of interactions characterizing it must underpin its longevity. Reversing the argument, this can be modelled by making large clusters less prone to collapse.

The particles released by a collapsed cluster are activated to move in real space by a unit step in a random direction. Model variants are characterized by different forms of $P(h)$. As it turns out, if $P(h)$ is integrable in $0 \le h \le \infty$, the equilibration time scale for the model dynamics diverges very quickly with increasing system size L and, in practice, the dynamics does not approach a stationary state even for rather small L values. In this ageing regime fewer and ever larger clusters gradually form and the MSD turns out to be similar to that of dense colloids. If, on the other hand, $P(h)$ is not integrable in $0 \le h \le \infty$, the dynamics quickly reaches a stationary state for any L. Here, clusters rapidly approach a constant average size and the particle's MSD is similar to that of diluted colloids. Since the *integrability* of $P(h)$ discriminates between the two types of behaviour Ref. [BS11b] considers the one parameter family of models given by

$$P^{\alpha}(h) = \frac{1}{\sum_{k=0}^{\alpha} \frac{h^k}{k!}}, \quad \alpha = 1, 2 \ldots. \tag{11.3}$$

Below we show results for $\alpha = 1$, where $P^{\alpha}(h)$ lacks integrability, and for $\alpha = \infty$, where $P(h) = \exp(-h)$ is integrable.

Figure (11.5) shows the model MSD for $\alpha = 1$ and $\alpha = \infty$. The former behaves diffusively in linear time (up to a cut-off scale $t_{st} \propto L^2$), and the latter diffusively in logarithmic time, as in the corresponding experimental result shown in Fig. (11.4).

As the collapse of large clusters controls all aspects of the model in the ageing regime, the statistics of such *quake* events is key to the analysis, a situation *identical* to ageing in quenched glasses [Sib07].

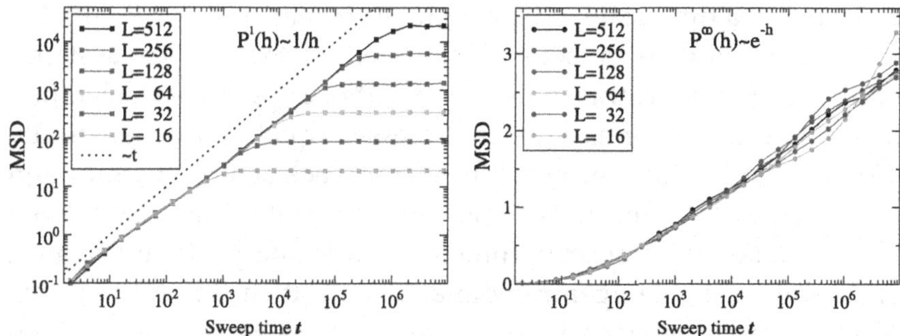

Figure 11.5. Both panels show the MSD obtained in model simulations versus time in units of MC sweeps. Several system sizes are included. Left: Normal diffusion obtained for cluster collapse probability $P(h) = \frac{1}{1+h}$. Right: Logarithmic diffusion obtained for $P(h) = \exp(-h)$. Figure taken from Ref. [BS11b].

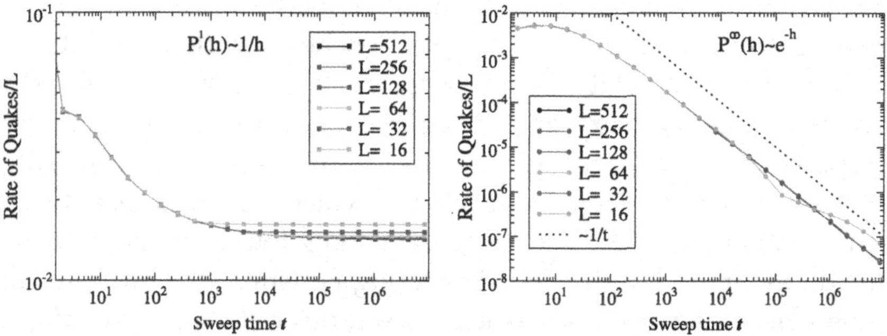

Figure 11.6. Both panels show the rate at which 'large' (defined in the main text) cluster collapse. Left: For the cluster collapse probability $P(h) = \frac{1}{1+h}$ the rate approaches a constant. Right: For $P(h) = \exp(-h)$ the rate is parallel to the dotted line and decays as $1/t$. Figure taken from Ref. [BS11b].

In turn, the fast collapse of small clusters provides a background of mobile particles diffusing within regions ('cages') bounded by large clusters. The divide between small and large clusters was empirically taken to be $\max(\bar{h}, \sigma(t))$, where $\sigma(t)$ is the (growing) standard deviation of the cluster size distribution. A quake is said to occur every time a cluster bigger than $\max\{\bar{h}, \sigma(t)\}$ breaks up. The quake rate versus time, scaled by the system size L, is shown in Fig. (11.6) for our simulations. For $P^1(h)$ the rate approaches a constant on a time scale independent of system size. For $P^{\infty}(h)$, the stationary regime is in fact invisible even for the smallest systems, and the dynamics is thus purely non-stationary with a decelerating quake rate decaying as $1/t$. As successive quakes are statistically independent events they provide the true clock of the dynamics, and the particle MSD simply becomes proportional to the time *integral* of the quake rate: Linear in time in the stationary (liquid) regime, and $\sim \log(t)$ in the ageing regime. Clearly then, the MSD in the ageing regime between times t_w and t scales as t/t_w, a so-called full ageing behaviour which is consistent with the experimental findings.

11.4. CONCLUSION

The dynamics of hard sphere colloids has a low- and a high density regime. In the former, the dynamics reaches a stationary state on experimentally available time scales, while in the latter the system enters a non-stationary ageing regime. The latter is in important respects similar to that of spin glasses. This is interesting considering that energy plays no role in the process, which must then be entropy driven. Dynamic light scattering data show that in dense colloids the decay of the Intermediate Scattering Function is logarithmic in time, and an analysis of particle trajectories obtained by confocal microscopy studies indicates that particle motion is diffusive when the data are parameterized by the logarithm of time. These findings are reproduced by a model of colloidal motions which relies on the destruction and reconstruction of patterns of kinematic constraints as the only mechanism allowing any net movement in the colloidal system.

12

Evolving Biological Systems

12.1. INTRODUCTION

Biology is rife with complexity, e.g. an ecosystem comprises many interacting and hierarchically organized components and offers a prime example of complex dynamics spanning many time scales. Three processes occurring in an ecosystem are easily identified: Development of organisms through their life cycle, sexual or asexual reproduction and mutation. The last two respectively control the relative abundance of different phenotypes, i.e. population dynamics, and the appearance of new types of organisms, i.e. evolution dynamics. Since the lifespan of individual organisms is typically much shorter than the characteristic time scale of both population and evolutionary dynamics, population dynamics is often treated as acting on a fixed set of types of organisms, and as leading the population to a local maximum of a fitness landscape, of which more later.

In the famous Lotka–Volterra model of a predator feeding on a prey, say foxes feeding on rabbits, rabbits and foxes can reasonably be assumed to reproduce without mutations. However, in the case of, e.g., humans, bacteria and viruses, the same assumption might not be accurate. During a human lifespan of, say 70 years, both bacteria and viruses reproduce so many times that mutations can change the relative frequencies of the phenotypes in human intestines. Finding effective antibiotics or vaccination is difficult precisely because both bacteria and viruses are able to mutate their way around the impediments posed by antibiotics or the raised immune defense stimulated by vaccination. The overlap of organisms' lifespans with the time scales of population and evolutionary dynamics is indeed common. Sequoias live for many centuries and other plants and microorganisms with reproduction times of the order of hours undergo in the meanwhile a great number of reproductive cycles.

A different question of importance is how an ecosystem responds to physical changes in its surroundings and whether such changes are crucial to its evolution or not, e.g. the fossil record exhibits long periods of low extinction activity separated by short time spans during which huge numbers of species abruptly disappear [Sim44, GE77]. This raises the issue of whether mass extinctions are mainly caused by external events such as meteorite impacts and volcano eruptions, or whether they can instead be caused by intrinsic instabilities of the highly interwoven dynamics of the earth's ecosystem. As seen in Chapter 8, complex systems tend to exhibit intermittency (quiescence interrupted by short hectic activity) as a result of their own internal interactions. This brings up the similarity between physics, the study of matter, and biology, the study of living systems, a theme implicit in the title of this book.

In the remainder of this chapter we discuss biological evolution models and highlight differences and similarities to the non-stationary complex dynamics seen in glassy systems. We emphasize the role of interactions between individuals and stress that mutations, selection and adaptation *change* the properties of ecosystems over many generations. By no means does our discussion intend to be a comprehensive account of mathematical models of evolution. The focus is instead on showing how complexity science can help to bring about new perspectives and insights in modelling biological evolution.

12.2. MODELS OF BIOLOGICAL EVOLUTION

From bacterial populations growing in a flask [LT94, BBL08] under controlled conditions to macroevolutionary processes imprinted in the palaeontological record [GE77], evolution can be studied at different levels of aggregation. Since individual organisms cannot function without strong couplings to other organisms, e.g. mammals and their bacterial flora, each organism is in a way similar to a whole ecosystem. Conversely, populations and species could be considered as individuals in a hierarchical theory of selection, a possibility discussed in depth by Stephen J. Gould [Gou02] in central chapters of his book *The Structure of Evolutionary Theory*. That common dynamical principles underlie processes occurring at different aggregation levels is a possibility which can be investigated *in silico* using computer models.

In the models discussed below, individuals are sketchily represented by bit strings or genomes and the rates of (asexual) reproduction rate and death are determined by their interactions combined with a stochastic element generically representing the influence of the environment. In spite of their simplicity, these models are able to reproduce key aspects of population and evolution dynamics, e.g. *punctuated equilibrium*, a name coined by Gould and Eldrege [GE77] in the early 1970s to describe the intermittent nature of the rate with which taxonomic structures undergo extinction and origination events. Punctuated equilibrium turned out to be controversial among biologists, perhaps because it seems to contradict the gradualism emphasized in Darwin's book *The Origin of Species*. It is now clear that whether changes are gradual or not strongly depends on the time scale and aggregation level of the description: Changes that are smooth and gradual at the level of individuals can lead to uneven and highly intermittent dynamics at the level of species and higher taxa.

We start with Eigen's quasi-species model, which looks at the formation of species as a dynamical process in sequence space. We then review fitness based evolutionary models, such as Kauffman's NK and NKC models. These models view evolution as an adaptive walk on fixed fitness landscapes. The Bak and Sneppen toy model is then briefly reviewed. Inspired by Self-Organized Criticality (SOC), this model was introduced to show that punctuated equilibrium arises naturally from interactions between species, a point of view we share even though we more strongly emphasize the non-stationary nature of evolution. Although the Bak–Sneppen model is really too simplistic to establish the separation of time scales it claims to explain, it did stir up a significant interest in biological evolution among people working in statistical mechanics and it certainly puts focus on the importance of the interaction between different species. Finally, we concentrate on the Tangled Nature model of evolutionary ecology. The model, which is defined at the level of interacting individuals avoids fitness as a dynamical variable, and explains the fate of individuals and the species to which they belong in terms of a web of interactions between the individuals of extant species. Intermittent events are triggered in the model by new mutants switching on yet unknown interactions patterns, possibly with large reorganization of the network of extant species as a consequence. The temporal sequence of such events is well described by a Poisson process with an average accumulated number of events depending on the logarithm of time, a fact which underlies the decelerating tempo of evolutionary dynamics already hinted to by the palaeontological record [NE99].

12.3. QUASI-SPECIES MODEL

The quasi-species model of Manfred Eigen and collaborators [EMS89] introduced reproducing sequences and fitness landscapes. The model is concerned with the temporal evolution of the relative frequencies $x_i = n_i/N$ of macromolecules, or sequences, where $n_i(t)$, for $i = 1, \ldots, M$, is the number of sequences of type i and $N = \sum_{i=1}^{M} n_i$ is the total number of sequences. When sequences reproduce, configurational changes or mutations allow a sequence of type j to turn into a sequence of a different type $i \neq j$. The rate of change in the frequency of i sequences due to reproduction, mutations and death is given by

$$\frac{dx_i}{dt} = \sum_{j=1}^{M} q_{ij}\, f_j x_j - \phi x_i. \tag{12.1}$$

Here f_j is the number of offspring per time unit a type j sequence can produce, i.e. in biological terms f_j is the reproductive or Malthusian fitness of a type j sequence. The matrix q_{ij} is the probability that one j sequence produces, due to mutations, an offspring of type i. Since the latter must be of one of the types $i = 1, \ldots, M$, $\sum_i q_{ij} = 1$ for each i. To ensure that the total number of individuals doesn't change, the parameter ϕ has the value

$$\phi = \sum_{i=1}^{n} f_i x_i. \tag{12.2}$$

With this choice of ϕ it is easy to check that $\sum_i \dot{x}_i = 0$, which ensures that the normalization $\sum_i x_i = 1$ is conserved by the dynamics. In the model, the fitness f_i of species x_i is an intrinsic property of type j, similar to a physiological characteristic, e.g. the number of legs.

Introduce now the column vector $\mathbf{x}$ and write Eq. (12.1) as

$$\frac{d\mathbf{x}}{dt} = \mathbf{W}\mathbf{x} - \phi\mathbf{x}, \tag{12.3}$$

where the elements of the matrix $\mathbf{W}$ are given by $W_{ij} = q_{ij} f_j$. The asymptotic composition of the population is determined by the properties of the eigenvalues of $\mathbf{W}$. If the mutation matrix $\mathbf{Q}$ is sufficiently 'narrow', i.e. if most of the offspring sequences are identical to their parent sequences, the population density in type space is concentrated in a set of well-defined peaks centered around a set of sequences called the wild types. But if flow between the different possible types, given by $\mathbf{Q}$, becomes too large, the population spreads out through type space. The transition between these two cases is called the error catastrophe and is an important general feature of evolutionary dynamics. We shall return to the issue in relation to the Tangled Nature model of evolution, see Section 12.6. We also notice that the dynamical equations for the quasi-species model are concerned with the *average* number of organisms x_i which belong to type i. In reality reproduction and mutation and death are all, at least to some extent, stochastic processes. As long as $x_i \gg 1$ neglecting stochastic fluctuations is unproblematic. When x_i becomes small, as will be the case near extinctions, fluctuations become important and one needs to go beyond the deterministic dynamics of Eqs. (12.1) and (12.3).

In conclusion, the quasi-species model focusses on the average flow between different types due to mutations but ignores that the presence of one type might influence the reproduction rate of another. Two different regimes exist depending on the frequency of mutations.

12.4. KAUFFMAN MODELS

In his book *At Home in the Universe*, Stuart Kauffman introduced two stochastic models, the so-called NK and NKC models, both based on fitness landscapes. The first deals with population dynamics and the second with the effects of competition between different species.

12.4.1. NK Model

In the NK model, individuals in a population are represented by their 'genome', i.e. by a string of N bits, the 'genes'. Each individual is therefore located at a point of the N-dimensional hypercube, a space on which one defines, in the way described further below, a one parameter family of real valued fitness functions F_K, where K is an integer with values in the interval K, $0 \leq K \leq N-1$. At each elementary step of a computer simulation, the number of individuals $n(x)$ present at site x is updated through reproduction, mutation and death: An individual reproduces with a probability proportional to $F_K(x)$. The function $F_K(x)$ signifies that the fitness of each of the N genes depends on the state of K other specified genes. Any offspring (or copy) undergoes (point) mutations with probability μ. Finally, individuals die with a fitness and site independent probability θ. As usual, the time unit in a simulation, i.e. the sweep, comprises a number of elementary steps equal to the number of individuals present in the system. Unlike the Eigen model, the NK model is fully stochastic, and provides an instance of an *adaptive random walk*. Since fitter individuals duplicate faster, the population clusters with high probability at the local fitness maxima of the landscape. Neglecting rapid transients, the dynamical behaviour can be coarse-grained into a series of movements of the whole population from one fitness peak to another. This process can be interpreted as the evolution of a species through successive stages of a lineage, with fitness becoming a property of a population of genetically identical individuals and hence of the whole species. The last feature is later used in connection with the NKC model.

The tunability of the fitness landscape of the NK model is expressed by the parameter K. The fitness $F_K(x)$ of genome $x = (x_1, x_2, \dots x_N)$, $x_i \in \{0, 1\}$ is defined by

$$F_K(x) = \frac{1}{N} \sum_{i=1}^{N} f_i(x, K), \tag{12.4}$$

where the contribution $f_i(x, K)$ from the ith gene and the K others it interacts with is a random function taking 2^{K+1} possible binary arguments and having values uniformly distributed in the interval $(0, 1]$. The coupling between different genes describes so-called *epistatic interactions*. If $K = 0$ a point mutation changes one of the N additive contributions to the fitness, and the change is of order $1/N$. The landscape is then called *smooth*. By way of contrast, when $K = N - 1$ one point mutation changes all the f_is, and hence changes the value of F_K to a new uncorrelated random number in the unit interval. In this case, the landscape is called *rugged*. Intermediate cases correspond, of course, to intermediate values of K.

In a rugged fitness landscape, each mutation gives the affected individual a new random fitness value. Most mutants will be misfits and quickly die out. To drive the population to a new fitness peak mutants need to have a fitness higher than the rest of the population. Neglecting that some favourable mutations could accidentally die out, *records* in the stationary series of random fitness numbers produced by successive mutations are precisely the events leading to macroscopic changes in population fitness. For $K = N - 1$, the evolution of the NK model in the hypercube is thus a simple instance of the record dynamics scenario described in Chapter 8.

The NK model can be modified by introducing phenotypes, i.e. sets of contiguous points sharing the same fitness value. Terracing the NK landscape in this fashion introduces neutral mutations which allow the system to move a long distance in genotype space without having to cross large areas of low fitness. As discussed in Ref. [SP99], the log-Poisson statistics of evolutionary events remains valid in terraced NK landscapes, and does so even for choices of K values significantly lower than $N - 1$.

Lenski and his group studied the evolution of *E. coli* bacterial colonies kept under constant physical conditions. Twelve different colonies started from genetically identical, i.e. cloned, bacteria are allowed to grow in an aqueous solution of nutrients which, initially, the bacteria have difficulties in metabolizing. Once the first phase of exponential bacterial growth has tapered off, a sample from each colony is injected in a flask of fresh solution, and a second generation of bacterial growth starts anew. At the time of writing, this process has continued for over 50000 generations. The (average) Malthusian fitness and cell size are measured for each generation, and both measures increase at a decelerating rate as a function of the generation number. The probabilistic nature of evolution is demonstrated by the fact that different bacterial colonies find different optimal ways to metabolize their nutrients, corresponding to different peaks in the NK fitness landscape. Qualitatively, these findings support the prediction of the NK model for population dynamics.

12.4.2. NKC Model

The NKC model considers the coarse-grained dynamics of co-evolving species, each characterized by a population which, at any time, is located at a local maximum of its fitness landscape. The fitness of species A is again a sum of contribution from different genes, or traits. However, each contribution now depends not only on K other genes of species A but also on C genes of species B, and vice versa. Depending on the relative values of K and C, qualitatively different behaviours can be observed. If K is large enough compared to C, the two species evolve independently. In the opposite limit, when the fitness of A increases, that of B may correspondingly decrease. In this case, B can more easily find a fitness peak better than its current one, and its rate of evolution is thus accelerated. Once B moves to a better fitness peak, it stimulates A to do the same. This evolutionary race, often termed the *red queen effect*, constitutes the so-called *chaotic* regime of the NKC model, a regime where the fitness changes erratically. The name notwithstanding, the regime is unrelated to chaos of non-linear dynamical systems with few degrees of freedom. Interestingly, the borderline between the frozen and the 'chaotic' regime, the *edge of chaos* is the parameter region where the average fitness of the system increases the fastest. It is not clear, though, how a species would optimize its couplings to other species in order to place itself at the edge of chaos.

12.5. BAK–SNEPPEN MODEL

If one accepts the evidence put forward by Gould and Eldredge for intermittency in evolutionary dynamics, the question remains whether its cause are external events, which themselves occur intermittently, or whether the property is inherent to evolutionary dynamics. We note that the two possibilities are not mutually exclusive and that the first choice merely relegates the problem to a different realm of science, e.g. planetary physics.

We are presently concerned with the aspects of the Bak–Sneppen (BS) model related to intermittency and punctuated equilibrium. For a careful discussion of the model from the more general view point of statistical mechanics and Self-Organized Criticality the comprehensive book by Pruessner [Pru11] should be consulted. The BS model is defined at the level of species [BS93] and seeks to demonstrate how intermittency can arise from species interdependence, i.e. the fact that the extinction of one species influences the chances for survival for other species interacting with it. However, by construction, the model neither offers an explanation of how intermittency at system level arises from smooth dynamics at the level of individuals, nor does it distinguish between evolutionary changes along a lineage and events where the lineage itself terminates.

The N 'species' of the model are each described by a single parameter $B_i \in [0, 1]$, the 'fitness barrier' of species number i, $i = 1, \ldots, N$. Species are organized in a ring, each species interacting with its two neighbours. At every time step the species i_{min} having the smallest fitness value is identified (by the All Seeing Eye) and removed from the list of extant species. The two contiguous species $i_{min} - 1$ and $i_{min} + 1$ are also dragged to extinction along with i_{min}, and similarly removed. To keep the number of species constant, three new species are then created having random fitness barriers $B_{i_{min-1}}$, $B_{i_{min}}$ and $B_{i_{min}+1}$. Iterating these dynamical steps asymptotically generates a probability density for the fitness barriers described by the step function

$$P(B) = \begin{cases} 1/(1 - B_c) & \text{if } B > B_c \\ 0 & \text{if } B < B_c, \end{cases} \tag{12.5}$$

where the threshold B_c is a parameter close to 2/3. So far, the dynamics is highly regular with exactly three species being renewed at every time step. BS introduces intermittency by postulating that the barrier height is a measure of the number of mutations needed in order to 'evolve' over the barrier. From this they suggest that the extinction time is related exponentially to the size of the barrier of the updated species. In other words, a single update in the computer algorithm involving a species with barrier $B_{i_{min}}$ is mapped into the time interval $\Delta t \propto \exp(B_{i_{min}}/T_{\text{char}})$, where T_{char} is some characteristic time scale pertinent to mutations. Through this blow-up mechanism, the smooth dynamics measured in terms of number of updates becomes highly intermittent because of fluctuations in the time values associated to the minimal barrier.

The model is patterned on the phenomenology for crossings of free energy barriers in glassy systems, but does not specify how a smooth regular pace of dynamics at the microscopic level changes into a highly intermittent and irregular tempo at the collective macroscopic level. In fact, since only three species are replaced at each event, a number which is negligible in the limit of large N, the distinction between micro- and macro-scales is moot in the model.

It is also important to point out the model reaches a stationary state. However, analysis of the fossil record points out that evolution is likely *not* to be a stationary process [NE99, Alr08].

12.6. TANGLED NATURE MODEL

Similarly to the Eigen and Kauffman NK models and to many other mathematical models of evolution, individuals are defined in the Tangled Nature Model by a bit string of length L, the length of their genome [CdCHJ02, AJ05, dCCJ03, HCdCJ02]. Individuals can reproduce and die, and their offspring is subject to point mutations, which produce a flow of individuals from one point of genome space, the L-dimensional hypercube, to one of its neighbouring points. Unlike, e.g., the NK model, the reproduction probability of individuals does not depend on a predefined fitness function but rather depends on the flow of external resources combined with the effect of a web of interactions which, over time, changes due to the change in the set of locations of extant species. The web of interactions is defined *in potentia* by a random matrix which may connect points of the hypercube. However, only the interactions involving extant species, i.e. points of genome space occupied by individuals at a given time, are of relevance at a given moment. This endows the model with two linked description levels: At the microscopic level and on short time scales, the number of individuals in occupied sites fluctuates reversibly, while at the macroscopic level and on much longer time scales the network of extant species evolves intermittently. This behaviour fully concurs with the idea of punctuated equilibria formulated by Gould and Eldredge [GE77], to describe the tempo and mode of macroevolution as inferred from palaeontological data. In the context of the Tangled Nature model, punctuated equilibria can be characterized statistically and appear to be the generic mode of the collective dynamics.

A slightly modified version of the Tangled Nature Model has been used by Rikvold and collaborators to study issues such as $1/f$ noise, stability and community structure and community assembly. See, e.g., Refs. [RZ03, Rik07, RS07, MSIR10].

12.6.1. Model Description

In the model, individuals (indexed by Greek letters e.g. α), are described in sequence space $\mathcal{S}$ by a binary string $\mathbf{S}^\alpha = (S_1^\alpha, S_2^\alpha, \ldots, S_L^\alpha)$, where all $S_i^\alpha = \pm 1$. In (asexual) reproduction a sequence duplicates itself, and while this happens, its components may mutate with probability p_{mut}, leading to the offspring $\mathbf{S}^\gamma$ having one or more components with a sign different from the mother's. The size of the type space is set by the length L of the sequences; a typical value used is $L = 20$ leading to about 10^6 different genotypes [CdCHJ02, AJ05, dCCJ03, HCdCJ02]. The sites in genome space represent all possible ways of constructing a 'genome'. While many sequences may not correspond to viable individuals, all possible sequences are available for evolution to select from.

A time step consists of *i*) an annihilation attempt followed by *ii*) a reproduction attempt. An annihilation attempt succeeds with a type independent probability p_{kill}. For simplicity, the killing probability is independent of the type and age of individuals. One generation consists of $N(t)/p_{\text{kill}}$ time steps, which is the number of time steps taken (on average) to kill all currently living individuals. When selected for reproduction, individuals of type $\mathbf{S}$ succeed with a probability $p_{\text{off}}(\mathbf{S}, t)$,

$$p_{\text{off}}(\mathbf{S}, t) = \frac{\exp[H(\mathbf{S}, t)]}{1 + \exp[H(\mathbf{S}, t)]}, \tag{12.6}$$

that depends on the sequence $\mathbf{S}$ and on the other types present in the type space through the weight function $H(\mathbf{S}, t)$. We note that the functional dependence described by Eq. (12.6) is chosen for convenience and that any other smooth differentiable monotonously increasing function mapping $H(\mathbf{S}, t)$ into the unit interval leads to similar dynamics. The weight function H contains the interactions which shape the dynamics and is given by

$$H(\mathbf{S}, t) = \frac{c}{N(t)} \left(\sum_{\mathbf{S}' \in \mathcal{S}} \mathbf{J}(\mathbf{S}, \mathbf{S}') n(\mathbf{S}', t) \right) - \mu N(t), \tag{12.7}$$

where c controls the strength of the interaction, $N(t)$ is the total number of individuals at time t, the sum is over the 2^L locations in $\mathcal{S}$ and $n(\mathbf{S},t)$ is the number of individuals occupying position $\mathbf{S}$. Two positions $\mathbf{S}^a$ and $\mathbf{S}^b$ in genome space are coupled or linked with a fixed strength $\mathbf{J}(\mathbf{S}^a,\mathbf{S}^b)$ which can be either positive, negative or zero. This link exists (in both directions) with probability θ, i.e. θ is simply the probability that the individuals residing on any two different sites are interacting. If the link exists, then $\mathbf{J}(\mathbf{S}^a,\mathbf{S}^b)$ and $\mathbf{J}(\mathbf{S}^b,\mathbf{S}^a)$ are drawn independently from a distribution which is symmetric in the interval $(-1,1)$. The parameter c fixes the overall range of the magnitudes of the interactions. The emphasis being on the effects of interactions *between* species, self-interactions are excluded by imposing $\mathbf{J}(\mathbf{S}^a,\mathbf{S}^a)=0$.

The effect of the physical environment enters in Eq. (12.7) through the term $\mu N(t)$, where the value of μ sets the average sustainable total population size, i.e. the carrying capacity of the environment. Assume that only one type $\mathbf{S}_0$ is present. Then $H=-\mu N(t)$ can be determined from the steady state condition

$$p_{\text{off}}(\mathbf{S}_0)=p_{\text{kill}}, \tag{12.8}$$

which using p_{off} given in Eq. (12.6) leads to $N(t) = -(1/\mu)\ln(1/p_{\text{off}}-1)$. An increase in μ thus reflects harsher physical conditions. Temporal variation in the environment, such as climate variation or sudden changes inflicted, say, by meteorite impacts or volcano eruptions, can be represented by allowing μ to be time dependent in a relevant way. We finally note that the total population the system can support is bounded from above: If N increases, the second term, $-\mu N$, eventually dominates the r.h.s. of Eq. (12.7), no matter what the links between extant types are. Since the offspring probability given by Eq. (12.6) correspondingly decreases, N cannot grow without bounds.

12.6.2. General Qualitative Behaviour

The first issue arising is that of initial conditions. As it turns out, these are not crucially important: Consider the two extreme cases where the entire population is initially *i*) placed at a single position in type space, or *ii*) is distributed on different randomly chosen positions, corresponding to a random collection of initial types. In the first case, there are by construction no interactions. In the second, the interactions between different types are typically unfavourable and the population of most sites dies out, only leaving a single position occupied. Once the population size falls to a level where $-\mu N(t)$ is sufficiently small to allow a non-vanishing rate of reproduction, mutations will allow some offspring to leave their initial position and move into the surrounding genotype space. Viable networks of extant populations gradually acquire a larger share of mutualistic interactions between their different types. For a given total population, the offspring probabilities are correspondingly higher than in the absence of mutualistic interactions, and the system as a whole can support a larger population.

On longer time scales, the dynamics features long periods of relatively stability (quasi-Evolutionary Stable Strategies or q-ESSs) (Fig. (12.1)) interrupted by brief spells of hectic activity (quakes) where the network of extant types is reorganized. Unlike the BS model, the intermittent macro-dynamics is typically not in a stationary state. When the value of the interaction parameter c is sufficiently high, e.g. $c = 10$, the transition rate between q-ESS

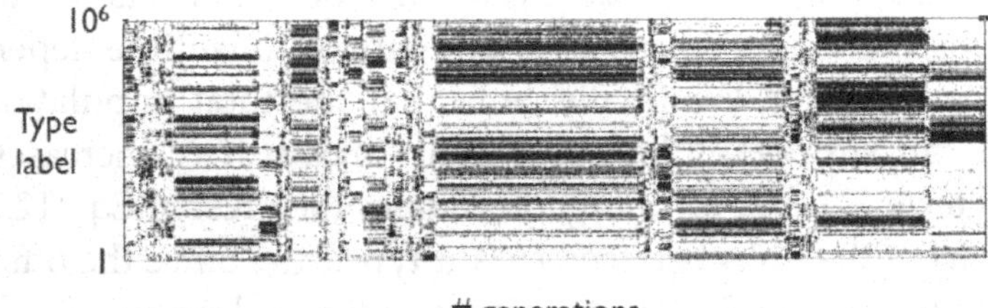

Figure 12.1. Intermittent evolution of the occupancy in type space. Time, measured in generations, is along the x-axis. The approximately 10^6 different types are labelled up along the y-axis. Whenever a type is occupied a dot is placed at its label. Long stretches of parallel lines indicate epochs during which the main composition in type space remains essentially the same. The figure courtesy of Matt Hall is from Ref. [HE10].

decreases with the age of the system [HCdCJ02]. Similarly to the quasi-species model, see Eq. (12.3), the Tangled Nature Model has an error threshold. When the mutation probability p_{mut} exceeds this threshold, the flow away from any occupied and reproducing type $\mathbf{S}^{\mathrm{occ}}$ becomes so large that the population persistently moves around between different possible types. Even if a particular well adjusted configuration is present at some instant in time, mutations will prevent the establishment of the large occupancy typical of q-ESSs. Fixation of a configuration is thus only possible when the reproduction of new offspring on a given location is larger than the outflow due to mutations and death [dCCJ03]. The following discussion of the model dynamics assumes mutation rates below the error threshold.

12.6.3. The Time Arrow in the Tangled Nature Model

The Tangled Nature model possesses two distinct temporal regimes: Time reversible fluctuations in population size characterize its dynamical behaviour on short time scales. On longer time scales the dynamics is intermittent, and large population changes of both signs can be observed in connection with the quakes. The probability that the population increase after a quake is slightly greater than that of a decrease (see Fig. (12.5) below) and averaging over a large ensemble

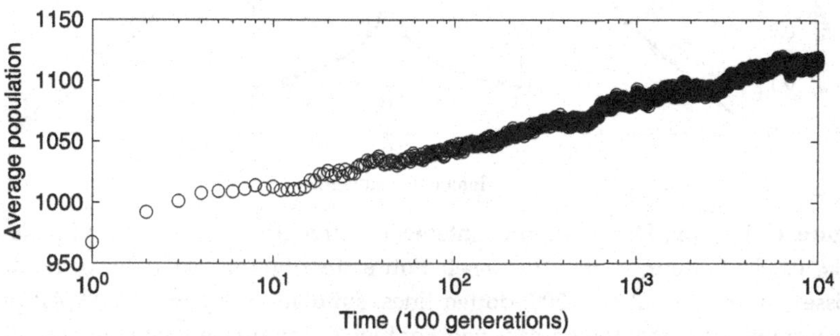

Figure 12.2. The dependence on time of the total population averaged over 1000 realizations. The figure is from Ref. [JJS10a].

of independent trajectories hence shows that the population on average increases, and that it does so in a logarithmic fashion. The increase of the average population $\langle N(t) \rangle$ shown in Fig. (12.2) comes about because the number of mutualistic interactions has a tendency to increase through the quakes, whence the first term in the weight function in Eq. (12.7) on average becomes more positive. Collectively then, the system becomes better adapted to the external environment (represented by μ) and gradually becomes able to support a larger population. Figure (12.3) shows the small but systematic increase in time of the number of mutualistic interactions between extant types. At the smallest connectivity value $\theta = 0.005$ the large shift observed is due to the viable part of the population being formed by isolated mutualistic pairs selected from pairs of positions with increasing

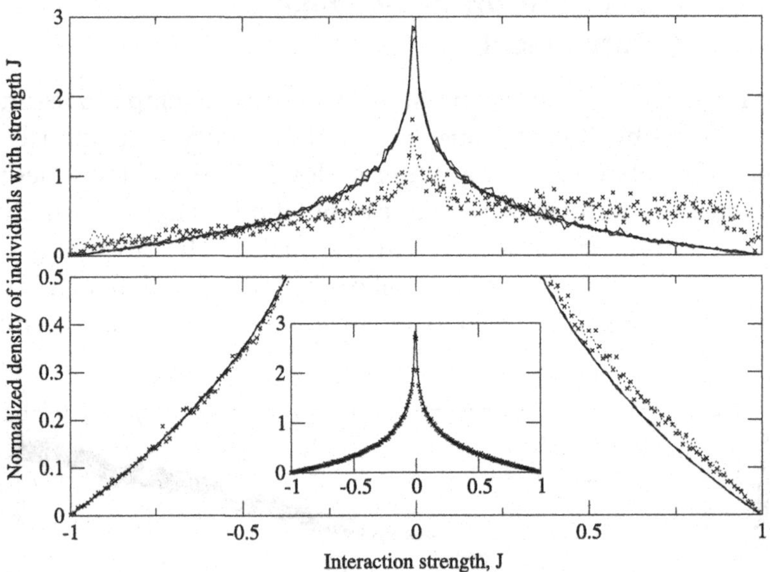

Figure 12.3. Top: Distribution of interaction strengths between individuals for $\theta = 0.005$. Bottom: $\theta = 0.25$. Inset: Entire distribution. Solid lines, random; crosses, simulation at t = 500; dotted lines, simulation at t = 500000. All plots are normalized so that their area is one. For high θ, a significant increase in positive interactions is seen. For low θ, a change is seen but for trivial reasons. The figure is from Ref. [AJ05].

interaction strength [AJ05]. For the larger connectivity value $\theta = 0.25$ the network of extant types forms one large connected structure. In this case one observes a significantly weaker shift towards more mutualistic interaction strength. This is due to the entire network collectively selecting a better distribution of interactions between populated sites. The temporal evolution of the total population show in Fig. (12.2) is relatively smooth due to averaging many different trajectories. However, in individual trajectories the intermittent transitions seen in Fig. (12.1) can be identified. On this basis, one may determine how the number of quakes increases on average with time. The left panel of Fig. (12.4) shows that on average this number (lower curve) is nearly proportional to the logarithm of time. Note that the linear dependence on log-time is consistent with the time dependence of $N(t)$ observed in Fig. (12.2). This reflects that the changes in the $J(\mathbf{S}^a, \mathbf{S}^b)$ couplings mainly occur during the quakes. The variance (upper curve) increases in the same fashion. In a log-Poisson distribution the two quantities would coincide. The discrepancy can be due to the inherent uncertainty in identifying the quakes leading to overcounting.

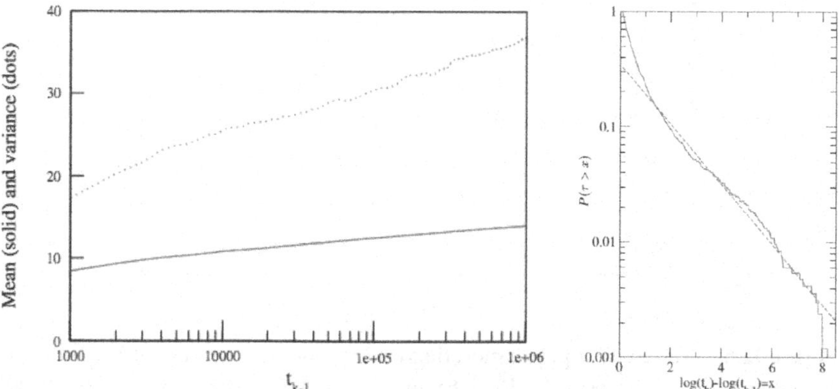

Figure 12.4. Left: The solid line indicates the accumulated number of transitions occurring on average up to time t. An estimate of the variance of the same quantity is shown as the dotted curve. Right: The cumulative distribution of log-waiting times $P(\tau > x)$.

The right panel of Fig. (12.4) shows the accumulated probability distribution $P(\tau > x)$ where τ is the logarithmic waiting time between successive events, i.e. $\tau_k = \log(t_k) - \log(t_{k-1})$ for $k = 1, 2 \ldots$. In log-Poisson statistics the distribution is exponential, which is near to the simulation results. To investigate in more detail the decay of the rate of quakes, we consider the fluctuations ΔN in the total populations size. These are defined as population changes occurring over a (short) time interval δt:

$$\Delta N(k) = N(t_w + k * \delta t) - N(t_w + (k-1)\delta t). \tag{12.9}$$

The fluctuations are sampled over 'narrow' time intervals of the form $(t_W, t_W + 4\delta t)$. In order to be able to extract the dependence of the statistics on the system age t_W, $\delta t = t_W/100$.

The plot in Fig. (12.5) shows PDFs pertaining to three different values of the system age t_W, each value ten times larger than its predecessor. The estimated PDF, $P(\Delta N)$, exhibits three interesting

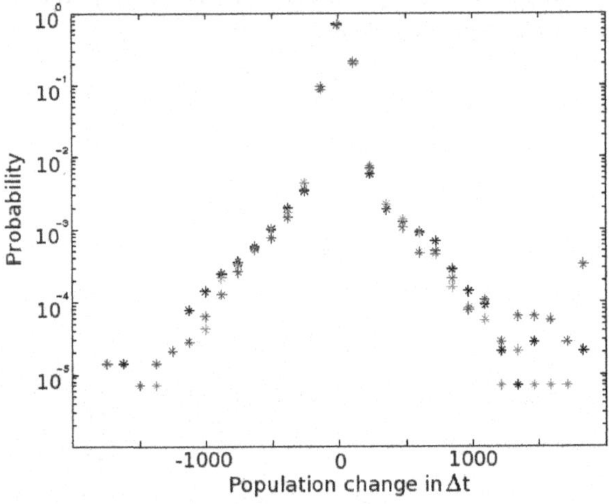

Figure 12.5. The PDF of population fluctuations for an ensemble of 1500 runs of 10^6 generations. Squares: $t_w = 1000$, triangles $t_w = 10000$, diamonds $t_w = 100000$. The total number of data points is equal for each age of the system; that is the sampling period $\Delta t = \frac{t_w}{100}$ scales with the age of the system. The data collapses onto a single curve, showing that the probability of large fluctuations scales like $\frac{1}{t_w}$. The figure is from Ref. [JJS10b].

properties. Firstly, small size fluctuations in the total population size are distributed according to the central Gaussian peak. Secondly, large fluctuations are described by exponential tails. Finally, the tails for the three different values of t_w nearly collapse on the same curve. Since the interval over which the change is calculated increases in proportion to t_w, the rate at which large changes occur must decrease as $1/t_w$. In contrast, the small Gaussian distributed fluctuations occur at a rate independent of system age: When increasing t_w with fixed Δt, the central peaks pertaining to different values of t_w still collapse on the same curve, whereas the exponential tails t_w are gradually suppressed, for details see [JJS10b].

We conclude that the statistics of the population size fluctuations and their age dependence strongly indicate that the transitions between the quiescent periods are controlled by record dynamics. However, so far it is not known which quantity undergoes the record fluctuations. In this respect, if the situation is very different from the spin glass case where it is well-established that the internal energy relaxes through record-sized fluctuations, see Chapter 9.

12.6.4. The Evolution of the Fluctuations in Birth Rate

It is interesting to relate the ageing processes observed in the Tangled Nature model and in models of spin glasses. First we consider Fig. (12.6), which shows the time evolution of the estimated standard deviation of the birth rate averaged over all populations [JJS10b]. To simplify the notation, we let the index i run over all the D different types for which $n(\mathbf{S}^{\mathrm{occ}}, t) > 0$ and write our estimated standard deviation as

$$\sigma_{p_{\mathrm{off}}} = \left[\frac{1}{D} \sum_{i=1}^{D} p_{\mathrm{off}}(i)^2 - \left(\frac{1}{D} \sum_{i=1}^{D} p_{\mathrm{off}}(i) \right)^2 \right]^{\frac{1}{2}}. \tag{12.10}$$

The data shown in Fig. (12.6) are sampled during ten time intervals $\Delta t_k = t_k^{(f)} - t_k^{(i)}$ spaced equidistantly along the time axis of the

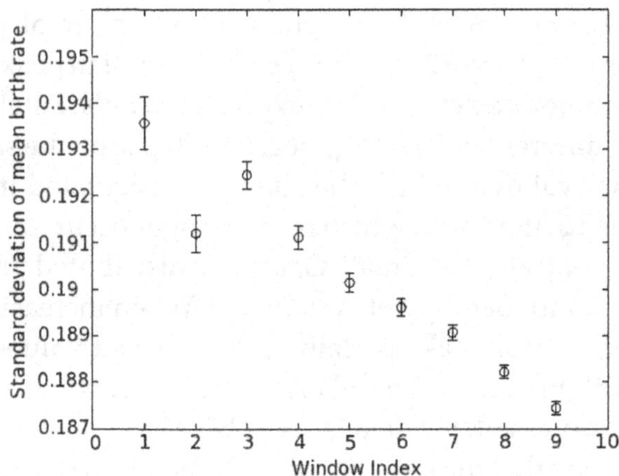

Figure 12.6. The standard deviation of birth rate fluctuations decreases over time. The window index, x, gives the start and end points of each sample window using the following conversion: $t_k^{(i)} = 10 \cdot 2^x$ and $t_k^{(f)} = 18 \cdot 2^x$. Error bars show that while the effect is small, it is significant. The figure is from Ref. [JJS10b].

systems evolution with $t_k^{(i)} = 10 \cdot 2^k$ and $t_k^{(f)} = 18 \cdot 2^k$. The standard deviation decreases logarithmically over time, and although the decrease is small, it is significant as it expresses the non-stationary nature of the dynamics. The decrease is related to the ageing of the system, which we now discuss. As described in Chapter 9, the thermally activated ageing dynamics of glassy systems can be coarse-grained into a barrier hopping process. When the system, through quakes, jumps from one metastable configuration to the next, the energy barrier characterizing successive configurations increases slightly. Consequently, jumps can be induced by record-sized energy fluctuations, and the dynamics decelerates as records in the thermal noise impinging on the system become increasingly rare.

In the Tangled Nature model, large values of the birthrate lead to a significant number of mutations. This enables the system to 'break out' of its quasi-equilibrium states as species mutate to nearby sites. As the fluctuations in the birth rate decreases, the 'bombardment' of sites adjacent to the network of occupied sites by new mutants also decreases, i.e. successful attempts to invade new territory in type space become increasingly improbable, and successive quasi-equilibrium states become thereby increasingly stable. This happens in conjunction with the network of interactions between extant types becoming more mutualistic, see Fig. (12.3), making the impact of a new mutant on the reproduction probability of the existing types relatively weaker. In an analogy to thermal hopping, this can perhaps correspond to a physical system jumping through energy barriers of constant size, but experiencing for each jump a small decrease in its temperature. Plausibly, a decreasing fluctuation range means that the adaptive dynamics takes the set of occupied types $\{\mathbf{S}^{\mathrm{occ}}\}$ into regions of type space surrounded by a periphery predominantly coupled to the rest of the system through couplings with negative ends pointing towards the mutant sites. Mutants entering these peripheral sites most likely die off. Hereby they are prevented from destabilizing the system by populating sites further afield.

12.6.5. The Offspring Probability and Adaptation

In his book *The Origin of Species*, Darwin argued that, because of the competition between all living organisms, the offspring of an individual which undergoes a beneficial mutation dominates future generations by outnumbering the offspring of less fortunate individuals. The Tangled Nature Model is simple enough to enable us to pinpoint how such direction or trend at the macroscopic level emerges from its undirected microdynamics.

Fluctuations in the weight function $H(\mathbf{S}^{\mathrm{occ}})$ given in Eq. (12.7) are due to mutations which change the occupancy in type space. By construction, these are symmetric, i.e., it is equally likely that a fluctuation occurs which drives $H(\mathbf{S}^{\mathrm{occ}})$ down by a certain amount $H(\mathbf{S}^{\mathrm{occ}}) \mapsto H_-(\mathbf{S}^{\mathrm{occ}}) = H(\mathbf{S}^{\mathrm{occ}}) - \Delta$ as it is that a fluctuation occurs which increases $H(\mathbf{S}^{\mathrm{occ}})$ by the same amount $H(\mathbf{S}^{\mathrm{occ}}) \mapsto H_+(\mathbf{S}^{\mathrm{occ}}) = H(\mathbf{S}^{\mathrm{occ}}) + \Delta$. The reason why the types undergoing changes to H_+ on average out number those subject to the change H_- is that the offspring probability function p_{off} in Eq. (12.6) is convex in the region where $p_{\mathrm{off}}(H) \approx p_{\mathrm{kill}}$, i.e. precisely in the region where the system finds itself during periods in which the population is on average constant. The convexity ensures that the increase in p_{off} induced by the fluctuation to H_+ is larger than the decrease induced by the fluctuation down to H_-. Or in other words

$$p_{\text{off}}(H_+) - p_{\text{off}}(H) > -[P_{\text{off}}(H_-) - p_{\text{off}}(H)]. \tag{12.11}$$

As a check, we can investigate the effect of using an offspring probabilility $p_{\text{off}}(H)$ without any convex regions, i.e. the piecewise linear function

$$P^*_{\text{off}}(H) \begin{cases} 0 & \text{if } H \leq -2 \\ \dfrac{H+2}{4} & \text{if } |H| < 2 \\ 1 & \text{if } H \geq 2. \end{cases} \tag{12.12}$$

When this is used, the system loses its ability to evolve to configurations able to support a larger number of individuals: Fig. (12.7) shows that, in contrast to the logarithmic increase seen in Fig. (12.2), $N(t)$ now fluctuates around a time independent average value. Which functional form for the offspring probability $p_{\text{off}}(H)$ is the more realistic? At first it might appear there isn't much difference between the functional shape of the function defined in Eq. (12.6) and the one defined in Eq. (12.12). But, on second thought, one will probably agree that in order to avoid any convex regions $p_{\text{off}}(H)$ must have some singularities. In the example in Eq. (12.12) the function $p_{\text{off}}(H)$

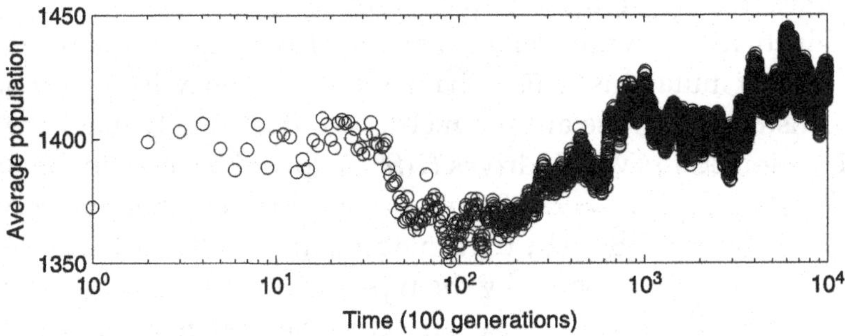

Figure 12.7. The same as Fig. (12.2), i.e. the dependence on time of the total population averaged over 1000 realizations, except the offspring function used (see Eq. (12.12)), doesn't have any convex regions. Figure courtesy of Dominic Jones.

is not differentiable at the point $H = \pm 2$. It is unclear how such singular behaviour could arise in a generic way.

12.7. CONCLUSIONS

As briefly discussed in this chapter, aspects of evolution can be studied using simple computer models which treat individuals as bitstrings in an abstract configuration space sometimes called a landscape. In some models, the reproductive success of individuals is determined by a (Malthusian) fitness function defined on the landscape. Such models can describe a single population evolving in a fixed environment, e.g. the bacterial evolution experiments of Lenski and Travisano. At any time, the population resides at or near a peak of the fitness landscape, and an evolutionary jump from one peak to another is triggered when a mutation leads to a fitness record. This is very similar to the ageing of a spin glass, as described by a single point – standing for the system configuration – moving in an energy landscape. An approach closer to real evolution is taken in the Tangled Nature model, where reproductive success of individuals is determined by the web of interactions connecting the species to which they belong and other extant species. In this way, it is possible to clearly identify two (spatial and temporal) levels of description: Individuals are born and die on short time scales, during which the web of extant species appears to be stable. On much longer time scales, the network of extant species changes, and does so at a decelerating rate, again following a statistics strongly reminiscent of record dynamics. The growing stability of the extant networks depends on the fact that the collective adaptive walk performed by the configuration in type space is able to bring the extant population into regions where the occupied sites are increasingly surrounded by a periphery of sites on which mutants have a decreasing probability of causing havoc. This scenario seems to be supported by the observation that the fluctuations in the reproduction rate decrease with system age. There is no energy or fitness function in the Tangled Nature model, but for the sake of comparison, a similar dynamics is achieved by a particle such as an interstitial defect which diffuses in a periodic potential, while the host crystal slowly cools down. Even though the energy barriers seen by the particle do not change in time, the rate of thermally activated hopping decreases with decreasing temperature.

Complex systems have many levels of description, rather than simply two. Indeed the distinction between individuals and systems is moot, or is, at least, predicated on a choice which may be convenient but remains arbitrary to a degree. Darwin insisted on evolution as a process involving random mutations of individuals, followed by a selection depending on both mutual interactions and interactions with the environment. Gould proposed a macroevolutionary theory emphasizing that different entities, e.g. species, can assume the role of individuals in a larger scale description. But how are random mutations then induced at the level of such 'collective' individuals?

Perhaps, an answer could be found exploiting the surprising similarities between complex dynamics of physical and biological origin. In the first case, a tiny increase in temperature imposed on, e.g., an ageing spin glass leads to rates of macroscopic change increasing to values typical of younger system ageing at constant temperature, whence the name 'rejuvenation' is used for the phenomenon. If this tiny increase of temperature is, after a while, followed by a decrease back to the previous value, the system falls back into a (collective) state statistically similar, but not identical, to the original state, i.e. the spin glass has been partly randomized at the *systemic* level, in contrast to the randomization of the directions of single spins which follows from random energy fluctuations and which happens on much shorter time and length scales.

Turning to biological systems, these are all the time subject to cyclic changes in external parameters which happen on a wide range of time scales and range from daily temperature and light changes to seasonal and even climatic changes. We suggest that a complex biological system possesses the same overall features as a spin glass, i.e. that, similarly to spin glasses, cyclic changes can induce partial randomizations at the systemic level, which, in biological terms, appear as mutations at the system level. It is not yet clear which parameters could play the same role as temperature does for the spin glass case. The Tangled Nature model has only one external parameter, μ, modelling resource availability. Assuming that the 'collective' mutation effects induced by cycling the value of μ can be identified, their effects on the interactions among large scale entities has, of course, to be described.

13

Non-stationary Ageing Dynamics in Ant Societies

13.1. INTRODUCTION

This chapter is concerned with a recent intriguing observation, namely that intermittent record dynamics is also of relevance for the collective behaviour of certain ant colonies. In their study on ant motion in *Temnothorax albipennis* ant colonies, Richardson *et al.* [RRC+10, RCF+11] found that ants leave their nest at a decelerating rate, and that the sequence of their exit times can be modelled using the log-Poisson process repeatedly encountered in this book. Our discussion closely follows Ref. [SC11], where a model is introduced to explain the ant exit statistics. The same model seems more generally applicable to describe aspects of the dynamics of interacting agents society.

The model is concerned with agents, i.e. ants, which move in a probabilistic fashion within a lattice with periodic boundary conditions, i.e. a torus. Contiguous points in the lattice each represent contiguous areas of physical space. Generally speaking, the motion of one ant is dependent on the position of all the other ants located within a certain distance from it. Allowed moves involve the ant changing its position from one lattice point to one of its four neighbours. Moves are accepted or rejected with probabilities depending on the ratio $\delta E/T$ between δE, the change they entail in a utility function E, and a parameter T called the Degree of Stochasticity, or DS. The symbols E and T convey that the roles of these parameters are similar to those of the energy and temperature of a physical system, i.e. if $T = 0$ the dynamics is deterministic, and all moves attempt to *maximize* (rather than minimize) the value of E. At high values of T, all moves are equally probable, and the agents diffuse within their confinement with no concern about the whereabouts of their fellows.

The strong similarity with, e.g., spin glasses and other glassy systems lies in the following property: At high enough T a stationary (equilibrium) regime is quickly attained. At low values of T, and depending on the structure of the interaction matrix, the system may not be able to reach a stationary regime. The dynamics is then decelerating but becomes time homogeneous if time is replaced with its logarithm. In this regime, the statistics of movements is well approximated by a log-Poisson process.

13.2. MODEL DESCRIPTION

The only property assigned to ants in the model is their 'type', which in turn determines the interaction patterns controlling the ant movements in the system. In other words, ants can move around, and in some case cluster in spatial patterns, but do nothing beyond that. There are N different types moving on a toroidal lattice of size $M = L^2$, where L is the linear grid size. Using typewriter order (the natural order of letters in a text written in a western language) the grid points are mapped into a one-dimensional array, and a configuration is specified by the number $n_{l,x}$ of individuals of type l located at site x, i.e. by the $N \times M$ rectangular matrix $\mathbf{n}$. The interactions J_{ij} between a pair of individuals of type j and i are ordered in the symmetric $N \times N$ square matrix J. Denoting matrix transposition by a superscripted dagger, (†), we construct the $M \times M$ matrix

$$\mathbf{I} = \mathbf{n}^\dagger \mathbf{J}\mathbf{n}. \tag{13.1}$$

Each entry,

$$I_{xy} = \sum_{k=1}^{N}\sum_{l=1}^{N} n_{k,x}\, J_{k,l}\, n_{l,y},$$

represents the interaction of all individuals at site x with all individuals at site y, irrespective of the Euclidean distance $d_{x,y}$ separating these two sites. To weigh distance in, a 'damping' matrix $\mathbf{D}$ is introduced with entries $D_{x,y} = \exp(-\alpha d_{x,y}^2)$, where α is a non-negative constant. Note that D is symmetric and has diagonal elements equal to one. The total interaction 'utility function' is finally defined by

$$E = \text{trace}(\mathbf{ID}) = \sum_{x=1}^{M}\sum_{y=1}^{M} D_{x,y} I_{y,x}. \tag{13.2}$$

In the equation, each term of the outer sum is the contribution to E stemming from site x. Corresponding to Eq. (13.2), the utility function change associated to the move of an individual of type l from site f(rom) to site t(o) is given by

$$dE_{l,t,f} = 2\sum_{x=1}^{M}\sum_{i=1}^{N}(D_{x,t} - D_{x,f})n_{i,x}J_{l,i} + 2J_{l,l}(1 - D_{t,f}). \tag{13.3}$$

If the system only contains a single ant of type l, i.e. if $n_{i,x} = \delta_{i,l}\delta_{x,f}$, the utility function change associated to any move from f to t is always zero. The same applies if the interactions are independent of distance, e.g. if D has all entries equal to one. In both cases, the motion is purely diffusive. To connect with ant experiments, an arbitrary site on the grid is designated 'exit', and all movements which involve this particular site are recorded as 'events'. Depending on the situation, we consider the statistics of either the waiting times $t_k - t_{k-1}$ between consecutive events, or that of the corresponding log-waiting times $\ln(t_k) - \ln(t_{k-1})$. Secondly, we calculate the time dependence of the average utility function for a number of different situations. Thirdly and finally, we show that, in the ageing regime, the probability of suitably defined 'large' events occurring in an interval $[t_w, t]$ anywhere on the grid scales as $\ln(t/t_w)$, modulo finite time corrections. As mentioned below, these are due to our time variable being restricted to integer values.

The dynamics starts out from a configuration where the number of ants of each type located at a site is drawn, independently for each site and type, from a uniform distribution between zero and ten. The steps make up a Markov chain generated by the Metropolis rule with the temperature T renamed 'degree of stochasticity' (DS). There is a minor adjustment in a sign convention, i.e. the utility function increases on average with decreasing DS, rather than decreasing with temperature as is the case for the energy in a physical system. In each update, a position is randomly chosen with equal probability among all those available. A type is then chosen in the same fashion. An ant of the given type (if present) is assigned a candidate move to a randomly chosen neighbouring site. The move is accepted if it increases the value of the utility function. Moves decreasing the utility function are accepted with a probability which decays exponentially as a function of the ratio $dE_{l,t,f}/T$ of the utility function change to the DS. For $T = 0$ moves decreasing the utility function are always rejected, and the dynamics is a greedy optimization algorithm attempting to maximize E. In equilibrium, the probability for configuration k is, modulo a normalization constant, equal to $\exp(E(k)/T)$. Whether the stationary distribution is within reach strongly depends on T, as further discussed below. Time evolution is gauged in terms of Monte Carlo (MC) sweeps, each sweep comprising a number of queries equal to the number of individuals in the system. We note from the outset that using a discrete time variable we cannot expect the exit time statistics to fully match the properties of a Poisson or log-Poisson process, where time is continuous.

13.3. GENERAL CONSIDERATIONS

Behind the decelerating nature of ageing dynamics is the gradual *entrenchment* of dynamical trajectories in more long-lived metastable configurations [SD03, PAS04]. In our model ants which attract each other tend to cluster. Starting from a random distribution, ever larger clusters get established on gradually fewer sites. As discussed below, this in turn creates growing dynamical barriers (e.g. empty sites) for ants which have not yet joined a cluster.

To further investigate this issue, several choices of interactions were considered. Our first choice is a null model lacking any metastable configurations. Ants of the same kind repel each other, while ants of different kinds attract each other. Correspondingly, the interaction matrix has diagonal and off-diagonal elements respectively equal to -1 and to 1. In this case, ants of the same type tend to maximize their mutual distance and hence to spread out uniformly in space, irrespective of type. Sites end up either being empty or being occupied by ants of different types. For *all* values of the DS, the dynamics quickly converges to a stationary state, where 'exit' events are (nearly) a standard Poisson process, of the sort illustrated (for a different example) in the left panel of Fig. (13.1). The mean utility function relaxes for all T's in a way similar to the curves at $T = 200$ and 500 which are shown in the left panel of Fig. (13.2).

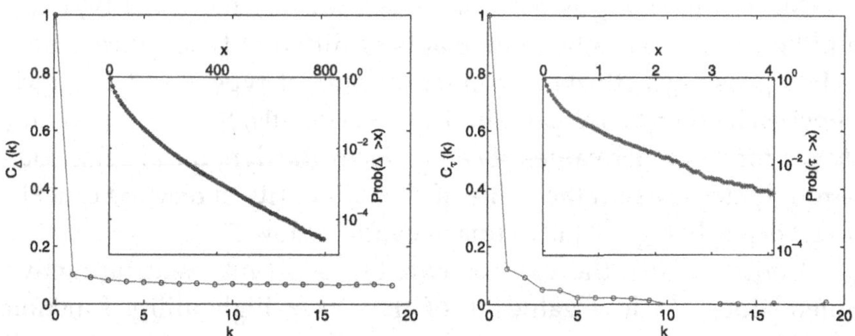

Figure 13.1. Left: The correlation function $C_\Delta(k)$ for the waiting times between consecutive exit events, plotted vs k. The insert shows the cumulative distribution of the waiting times, plotted on a logarithmic vertical scale. The system contains two types of ants moving at $T = 50$ on a grid of linear size 7. Right: Same as above, except that $T = 5$ and that the correlation and cumulative distribution are calculated using the log-waiting times rather than the waiting times. Figure taken from Ref. [SC11].

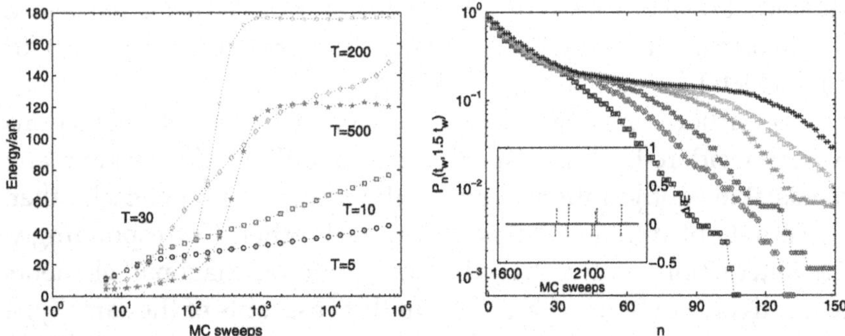

Figure 13.2. All data shown pertain to a system with four ant types, with one negative and five positive interactions between different types. Left panel: The utility function per ant, averaged over 100 trajectories, is plotted versus time. The initial configuration is in all cases obtained by randomly placing the ants on the grid. Right panel: The probability that the number of 'large' events (as defined in the main text) occurring in the interval $[t_w, 3/2t_w]$ is larger or equal to n is estimated using 2500 trajectories all run at $T = 10$. The results are plotted versus n for sampling intervals starting at $t_w = 800$ (lowest curve) and ending at $t_w = 25400$ (highest curve), via intermediate values $t_w = 1600, 3200, 6400$ and 12800. The insert shows a sample trajectory starting at $t_w = 1600$. Figure taken from Ref. [SC11].

More interesting is the case where ants of the same type are indifferent to each other, while ants of different types attract each other, possibly with the exception of ants of type 1 and 2, which repel each other. Correspondingly, $J_{ii} = 0$ for all i, $J_{12} = J_{21} = -1$ and $J_{ij} = 1$ for any other values of i and j. Here, the dynamical behaviour strongly depends on the value of T: Stationarity is reached quickly at sufficiently high T but is unachievable at low T.

Consider first the case of two types of ants with attractive interactions. At low values T of the DS, a high utility function equilibrium configuration has the two types grouped into a small number of sites and, eventually, into just a single site. Starting from a random initial distribution, empty areas gradually form. Since two metastable clusters located at different sites can increase their utility function by merging, a process which requires crossing empty areas of growing size, the model features growing dynamical barriers between metastable configurations. As shown in Fig. (13.1), this model version has Poisson and log-Poisson exit statistics at high, respectively low T.

Consider now four types, with pairs $(1-2)$ repelling each other, and all other pairs attracting each other. Spatially segregated domains of type 1 and 2 must gradually form, while groups of ants of type $3-4$, each centered on a particular site, form within each of the two domains. Besides empty space, a domain d_2 containing ants of type 2 presents an additional dynamical barrier for ants of type 1, since it either must be crossed or circumvented in order for these to join a domain d_1 located on the opposite side of d_2. Also in this case, low T ageing behaviour is found to characterize the dynamics.

13.4. DATA ANALYSIS

In all our simulations, the damping parameter α is equal to five, i.e. interactions are strongly localized in space. The spatially averaged density of each type of ant is close to 1.8, e.g. with four types present, the system contains 7.2 ants per grid site. Linear grid sizes $L = 5, 7$ and 9 were investigated for four different ant systems and seen to have qualitatively similar behaviours. Linear grid size $L = 7$ was used for systems with $2, 3, 4$ and 5 ant types. The results shown are for $L = 7$ to emphasize that a large grid size is not required to obtain ageing behaviour.

Figure (13.1) describes the exit statistics in a system where two types of ants are present. At the k'th sweep the program checks whether motion has occurred at the site dubbed 'exit' and, if so, registers the corresponding time t_k. The simulations, each running from $t = 5$ to $t = 5 \cdot (1 + 10^5)$, are repeated 100 times in order to improve the statistics. Simulations are performed at $T = 50$ and $T = 5$ in the left and right panel, respectively. The data shown are statistically very different in spite of being graphically rather similar. In the left panel the waiting times, i.e. the time differences $\Delta_k = t_k - t_{k-1}$ are analysed with respect to their correlation and their distribution. Since the t_k' s are integers, the exit process can never be truly Poissonian. We nevertheless estimate the normalized correlation function of the Δ_k's, averaged over 100 independent runs. For independent entries, the latter would equal the Kronecker's delta $C_\Delta(k) = \delta_{k,0}$. We furthermore estimate the probability that $\Delta > x$, as a function of x. For a Poisson process this probability decays exponentially in x. The correlation and probability distribution are plotted in the main figure and the insert, using a linear and a logarithmic ordinate, respectively. We see that short waiting times, i.e. Δ_k's of order one are over-represented relative to the straight line representing the Poisson case. Secondly, the correlation decays to about 1/10 in a single step, but then lingers at that value. Taken together, these two feature indicate that a short waiting time is more likely followed by another short waiting time, i.e. that the motion often stretches over several sweeps.

The right panel of the figure shows data obtained as just discussed, except that logarithmic time differences $\tau_k = \ln(t_k) - \ln(t_{k-1})$ rather than linear ones are utilized. The correlation function $C_\tau(k)$ decays quickly to zero, albeit not in a single step, and the probability that $\tau > x$ is nearly exponential. Again, short log-waiting times are over-represented in the distribution, and since the correlation decays to near zero in $k = 5$, they are likely to follow each other. Thus, also in this case an ant's 'exit' stretches beyond a single sweep. In summary, banning the effect of our time unit, the sweep, being too short relative to the decorrelation time of ant motion, $T = 50$ data are, as expected in a stationary regime, well described by a Poisson process, while $T = 5$ data are well described by a log-Poisson process.

The average utility function plotted vs time provides a good overall characterization of the model dynamics. At low T, the statistics of utility function fluctuations provides further insight on the nature of ant motion. At the level of a single trajectory, E appears on average constant, i.e. defining fluctuations δE as differences between consecutive sweeps, most of the time $\delta E = 0$. Hence, ants do not move outside the area represented by a single point. Occasionally, a negative fluctuation is quickly followed by one or several positive fluctuations. Such events indicate motion across a barrier from one metastable configuration to another, the latter usually having a higher value of E. Positive fluctuations larger than the threshold value $\delta E = 0.22$ are defined as large events, irrespective of where they occur in the system.

At low temperatures, E grows on average logarithmically in time, see the left panel of Fig. (13.2), and the typical number of large fluctuations can hence be expected to do the same. To study how this number actually behaves, the probability $P_n(t_w, \frac{3}{2}t_w)$, shown in the right panel of the same figure, that at least n 'large' events occur in the observation time interval $[t_w, \frac{3}{2}t_w]$ is estimated based on 2500 independent trajectories. Using logarithmically equidistant values of t_w, i.e. $t_w = 800 \cdot 2^k$, with $k = 0, 1, \ldots 5$, the observation time interval is doubled for each successive value of k. Since the difference between the logarithm of the respective endpoints remains equal to $\ln(3/2)$, the graphs of $P_n(t_w, \frac{3}{2}t_w)$ versus n corresponding to different t_w's would collapse if these functions only depended on the logarithm of time. Data collapse is present for values of n small compared to the length of the observation interval. At larger n, deviations arise since no more than one event is registered per sweep and the distribution hence explicitly depends on the length of the observation interval. In other words, the finiteness of the latter introduces corrections to log-time scaling.

All data presented in Fig. (13.2) pertain to a system with four types of ants. Its left panel depicts on a logarithmic horizontal scale the utility function per ant, averaged over 100, trajectories, as a function of time (i.e. number of MC sweeps). For $T = 500$ and 200, a constant value is approached, which, as expected, increases as the DS decreases. For $T = 5$ and $T = 10$ the mean utility function is seen to grow logarithmically without approaching equilibrium. The mean utility function value in equilibrium would lie far above the plateau reached at $T = 200$. If the logarithmic trend were to continue up to its asymptotic long time limit, the 'transient' would stretch over more than ten decades. The behaviour at $T = 30$ is intermediate, between the low and high T cases. Here, the curve has two knees, with a third barely visible at the end of the range. Yet, no stationary state is reached within the observation time. Consider now the two low T curves. For any fixed time value, the value of the mean utility function is higher the higher the degree of stochasticity. The strong non-equilibrium character of the low T dynamics is highlighted by the ordering being opposite rather than in equilibrium.

The statistics of fluctuation occurring at low values of the DS was investigated for $T = 5$ and $T = 10$, with data collected the latter case shown in the right panel of the figure. The insert illustrates the highly intermittent nature of the fluctuations. In the main panel, the probability of large events $P_n(t_w, \frac{3}{2}t_w]$ is plotted versus n for six different values of t_w, each twice as large as its predecessor. In the statistics, positive utility function fluctuation larger than $\Delta E_{tr} = 0.22$ are counted as large events, irrespective of where they occur in the system. The threshold is chosen to filter out small fluctuations which could be reversed, and the overall shape of the results is, within bounds, insensitive to the specific choice of threshold. Data sets belonging to different observation intervals collapse for values of the abscissa, n, smaller than the length of the observation interval. Banning finite time effects, the global dynamics is thus well described by a process which is homogeneous in the logarithm of time. This fully agrees with the statistical properties of the exit process which describes motion at a single site, but which, on the other hand is observed over a much longer time span.

We finally note that the same scaling analysis performed using data sampled at $T = 5$ yields a near perfect data collapse, with the sole exception of data taken during the shortest and earliest observation interval [800, 1200].

13.5. DISCUSSION

Following the seminal work of Axelrod and Hamilton [AH81] on the emergence of cooperative behaviour, probabilistic models and concepts rooted in game theory and equilibrium statistical physics [CFL09] are often applied to systems of interacting agents of biological or social interest, see, e.g., [Axe97]. The model discussed in this chapter strongly emphasizes non-equilibrium aspects of the dynamics of social agents. Indeed, equilibrium or steady state is *de facto* not achievable for a range of the model parameters, and the kinship of its dynamics to ageing of glassy systems clearly emerges, illustrating in a new context the general theme of this book.

The model deals with interacting agents (ants) which are able to move in a confined space. The model has different dynamical behaviours, mainly depending on the value of a parameter called the degree of stochasticity. The latter gauges the willingness of the ants to perform moves which decrease their utility function, i.e. their willingness to invest part of their resources, represented by a utility function, for the sake of future gains. At low DS values, and for interactions leading to a large number of metastable configurations, which are spatially localized clusters of ants, the model reproduces the experimental findings of Richardson *et al.* [RCF$^+$11]. At high DS values, the statistics of exits from the nest is a simple Poisson process, which is similar to the findings of Nouvellet *et al.* [NBW10].

The global pattern of ant movements is inferred from large utility function changes occurring anywhere in the system. The pattern is statistically described by the probability that n such events fall in a certain observation time interval. By considering a series of observation intervals of increasing duration, we ascertain that in the ageing regime the probability of large events scales with the difference of the logarithms of the interval endpoints, with finite time corrections most important for short observation intervals. In conclusion, at low DS values the dynamics is inhomogeneous and decelerating when parameterized by the usual time variable, but is to a good approximation homogeneous when parameterized by the logarithm of time.

The distinction between time homogeneous and non-homogeneous dynamics is of great conceptual importance. The first type reflects reversible fluctuations of observable quantities around fixed plateau values, while the second both implies the presence of a definite trend, and of a persistent memory of the initial condition.

That biological and physical systems may feature similar decelerating dynamics and that the latter is likely connected to record-breaking fluctuations is a point already made by Richardson *et al.* [RCF$^+$11] for ant motion, and previously made in different connections [SSA95, SBA98, NE99, PAS04, KJ05]. Along these lines, this model suggests a novel interpretation of record dynamics in the context of socially interacting agents: The record-sized fluctuations triggering important and irreversible dynamical changes correspond to record high investments, paid by a decrease of the utility function of the agents involved. This utility function represents the degree to which some condition, which is desirable for the agents to achieve, is actually achieved at each stage of the system evolution. What, more precisely, this function could be would, of course, depend on the application considered. However, the complexity of the dynamics reflects in all cases the presence of frustrated interactions, i.e. the fact that not all the agents can simultaneously maximize their utility function.

It would be interesting to experimentally identify what the 'utility function' and 'degree of stochasticity' are in a society of interacting agents, and to see whether such systems tune themselves to a DS value just below the threshold at which ageing disappears. This would give the fastest possible increase of the utility function compatible with keeping the memory and history dependence which seem natural in an evolving biological and social system.

Part III

Epilogue

14

What is Complexity Science?

About four decades ago, P.W. Anderson [And72] wrote a now famous paper entitled "More is different. Broken symmetry and the hierarchical nature of science". The thoughts he developed there are a good starting point to discuss what Complexity Science is today.

First and foremost, Complexity Science has a role to play because, in the hierarchy starting from particle physics and ending with social sciences, through solid state physics, chemistry and biology, each discipline has its emergent phenomena. These cannot be dealt with by blind mathematical application of the 'laws' which belong to a lower and allegedly more fundamental level of description. Instead, the phenomena must be identified and characterized at each level through an inductive process requiring specific tools. In Anderson's own words, "The ability to reduce everything to simple fundamental laws does not imply the ability to start from these laws and reconstruct the universe."

A prime example that 'more is different' and that a purely reductionist approach is inadequate, is broken symmetry. Macroscopic systems going through a second order phase transition spontaneously organise in a state whose symmetry is lower than the symmetry of the Hamiltonian describing the interactions, i.e. at a temperature below the critical point a magnetic material acquires a macroscopic dipole moment. Importantly, the orientation of the latter can in no way be predicted from the state of the material above the critical point. Hence, broken symmetry implies a loss of (deterministic) predictability, a feature shared by complex systems of all kinds. Underneath lies a partial blurring of the boundaries between different levels of a descriptive hierarchy, e.g. systems at their critical point lack any finite length or time scale and cannot be subdivided into spatially separated parts which are statistically independent of each other.

Critical phenomena are for many a powerful paradigm with strong implications for complex system science. In his book *How Nature Works. The Science of Self-Organized Criticality*, Per Bak [Bak97] argued that natural systems tune themselves into a critical state lacking a finite characteristic length scale. Self-Organized Criticality eliminates the far too restrictive need to tune a parameter to a critical value, but insists on the stationary state of a driven system as a model of complex behaviour. Yet biological evolution has, over the eons, generated a bewildering array of new structures in a process which can hardly be treated as stationary.

Emergence as discussed by P.W. Anderson is a collective property of systems in thermal equilibrium. It is possible, and indeed desirable, to focus on complex structures which appear permanent on a certain time scale, and equilibrium or steady state approaches as Self-Organized Criticality are useful to this end. In order to deal with the evolution of, e.g., life, the concept of emergence must however be extended to include the time domain. At a certain level of description, which we now for convenience shall call macroscopic, new structures appear and disappear in an intermittent fashion and for reasons hidden in microscopic details beyond our knowledge and control. Hence, knowing the full history of a system's evolution at the macroscopic level says only a little about future developments at the same level. This disquieting feature, which has been called 'ontic openness' [JPS99] in ecological modelling, also appears in human history where, to name just a few recent events, nobody predicted the demise of the Soviet Union, the collapse of Lehman Brothers Inc. or the explosive growth of the internet. In our view, ontic openness is rooted in the weak interdependence of nested description levels which, in the long run, shapes complex systems' dynamics through a series of unforeseen and disruptive events.

To forge mathematical tools with predictive power, a hierarchy of causal relationships below the desired description level must be closed in a self-consistent manner. In statistical physics, this device is used, for example, to obtain the Boltzmann equation by brute force truncation of the so-called Bogoliubov–Born–Green–Kirkwood or BBGK hierarchy. The point is that once we collectively realize that the Boltzmann equation actually works, a useful description of a dense gas — which happens to represent the macroscopic level in question — has emerged in our thinking.

In complex system dynamics, this type of procedure can at times lead to heuristic principles or statements with a tinge of circularity, such as the deprecated 'survival of the fittest', or any version of 'when the going gets tough, the tough get going'. Clearly, then, dealing with complex systems requires the combined insights of, e.g., biologists, physicists, sociologists, neuroscientists, economists etc. Thus, to fully develop the content and potential of Complexity Science, it is crucial to establish interdisciplinary collaborations based on a shared mathematical foundation.

To conclude: Can more be added to the concise characterization then given by Anderson of what complexity was forty years ago? We suggest, "Later is different. The gradual formation and sudden demise of ordered structures."

Bibliography

[AH81] R. Axelrod and W.D. Hamilton. The evolution of cooperation. *Science,* 211:1390–1396, 1981.

[AHOR87] M. Alba, J. Hammann, M. Ocio, and P. Refregier. Spin-glass dynamics from magnetic noise, relaxation, and susceptibility. *J. of Appl. Phys.,* 61:3683–3688, 1987.

[AJ05] P.E. Anderson and H.J. Jensen. Network properties, species abundance and evolution in a model of evolutionary ecology. *J. Theor. Biol.,* 232:551–558, 2005.

[Alr08] J. Alroy. Dynamics of origination and extinction in the marine fossil record. *Proc. Natl. Acad. Sci. USA,* 105:11536–11542, 2008.

[And72] P.W. Anderson. More is different. *Science,* 177:4047, 1972.

[AOH86] M. Alba, M. Ocio, and J. Hammann. Ageing Process and Response Function in Spin Glasses: An Analysis of the Thermoremanent Magnetization Decay in Ag:Mn(2.6%). *Europhys. Lett.,* 2:45–52, 1986.

[Axe97] R. Axelrod. The dissemination of culture. *Journal of Conflict Resolution,* 41:203–226, 1997.

[Bak97] P. Bak. *How Nature Works. The Science of Self-organized Criticality.* Oxford University Press, 1997.

[BBC03] L. Buisson, L. Bellon, and S. Ciliberto. Intermittency in aging. *J. Phys. Condens. Matter.,* 15:S1163, 2003.

[BBL08] Z.D. Blount, C.Z. Borland, and R.E. Lenski. Historical contingency and the evolution of a key innovation in an experimental population of *Escherichia coli. PNAS,* 105:7899, 2008.

[Ber10] R.S. Berry. Energy landscapes: topographies, interparticle forces and dynamics, and how they are related. *Theoretical Chemistry Accounts: Theory, Computation, and Modeling (Theoretica Chimica Acta),* 127:203–209, 2010.

[BF99] J. Bergenholtz and M. Fuchs. Nonergodicity transitions in colloidal suspensions with attractive interactions. *Phys. Rev. E*, 59:5706–5715, 1999.

[BH92] K. Binder and D.W. Heermann. *Monte Carlo Simulation in Statistical Physics: An Introduction*. Springer Verlag, Heidelberg, 1992.

[BH97] K. Binder and D.W. Heermann. *Monte Carlo Simulation in Statistical Mechanics*. Springer Series in Solid State Sciences, 1997.

[BK97] O.M. Becker and M. Karplus. The topology of multidimensional energy surfaces: Theory and application to peptide structure and kinetics. *J. Chem. Phys.*, 106:1495–1517, 1997.

[BKL75] A.B. Bortz, M.H. Kalos, and J.L. Lebowitz. A New Algorithm for Monte Carlo Simulation of Ising Spin Systems. *J. Comp. Phys.*, 17:10–18, 1975.

[BM86] J.G. Bednorz and K.A. Mueller. Possible high TC superconductivity in the Ba-La-Cu-O system. *Z. Phys. B*, 64:189–193, 1986.

[BM87] A.J. Bray and M.A. Moore. Chaotic nature of the spin-glass phase. *Phys. Rev. Lett.*, 58:57–60, 1987.

[BP98] K.E. Bassler and M. Paczuski. Simple model of superconducting vortex avalanches. *Phys. Rev. Lett.*, 81:3761–3764, 1998.

[BP01] S. Boettcher and A. Percus. Optimization with extremal dynamics. *Phys. Rev. Lett.*, 86:5211–5214, 2001.

[BPR99] K.E. Bassler, M. Paczuski, and G.F. Reiter. Braided rivers and superconducting vortex avalanches. *Phys. Rev. Lett.*, 83:3956–3959, 1999.

[BS93] P. Bak and K. Sneppen. Punctuated equilibrium and criticality in a simple model of evolution. *Phys. Rev. Lett.*, 71:4084–4087, 1993.

[BS11a] D. Barettin and P. Sibani. Entropic algorithms and the lid method as exploration tools for complex landscapes. *Phys. Rev. E*, 84:036706, 2011.

[BS11b] S. Boettcher and P. Sibani. Subdiffusion and intermittent dynamic fluctuations in the aging regime of concentrated hard spheres. *J. Phys. Condens. Matter*, 23:065103, 2011.

[BTW87] P. Bak, C. Tang, and K. Wiesenfeld. Self-Organized Criticality: An Explanation of 1/f Noise. *Phys. Rev. Lett.*, 59:381–384, 1987.

[Bur92] G. Burns. *High-Temperature Superconductivity: An Introduction*. Academic Press, San Diego, London, 1992.

[BY88] R.N. Bhatt and A.P. Young. Numerical studies of Ising spin glasses in two, three and four dimensions. *Phys. Rev. B*, 37:5606, 1988.

[CCF$^+$04] F. Cordero, F. Craciun, A. Franco, D. Piazza, and C. Galassi. Memory of Multiple Aging Stages above the Freezing Temperature in the Relaxor Ferroelectric PLZT. *Phys. Rev. Lett.*, 93(9):097601, 2004.

[CCFG10] F. Cordero, F. Craciun, A. Franco, and C. Galassi. Ageing and Memory in PLZT Above the Polar Freezing Temperature. *Ferroelectrics*, 319:19–26, 2010.

[CCW06] G.C. Cianci, R.E. Courtland, and E.R. Weeks. Correlations of structure and dynamics in an aging colloidal glass. *Solid State Communications*, 139(11–12):599–604, 2006.

[CdCHJ02] K. Christensen, S.A. di Collobiano, M. Hall, and H.J. Jensen. Tangled nature: A model of evolutionary ecology. *J. Theor. Biol.*, 216:73–84, 2002.

[CFL09] C. Castellano, S. Fortunato, and V. Loreto. Statistical physics of social dynamics. *Rev. Mod. Phys.*, 81:591–646, 2009.

[CKP97] L.F. Cugliandolo, J. Kurchan, and L. Peliti. Energy flow, partial equilibration, and effective temperature in systems with slow dynamics. *Phys. Rev. E*, 55:3898–3914, 1997.

[CMM$^+$90] L. Civale, A.D. Marwick, M.W. McElfresh, T.K. Worthington, A.P. Malozemoff, F.H. Holtzberg, J.R. Thompson, and M.A. Kirk. Defect independence of the irreversibility line in proton-irradiated Y-Ba-Cu-O crystals. *Phys. Rev. Lett.*, 65(9):1164–1167, 1990.

[CW03] R.E. Courtland and E.R. Weeks. Direct visualization of ageing in colloidal glasses. *J. Phys. Condens. Matter*, 15:S359–S365, 2003.

[dAT78] J.R.L. de Almeida and D.J. Thouless. Stability of the Sherrington-Kirkpatrick solution of a spin glass model. *J. Phys. A*, 11:983–990, 1978.

[dCCJ03] S.A. di Collobiano, K. Christensen, and H.J. Jensen. The tangled nature model as an evolving quasi-species model. *J. Phys. A*, 36:883–891, 2003.

[Dob73] T. Dobzhansky. Nothing in biology makes sense except in the light of evolution. *The American Biology Teacher*, 35:125–129, 1973.

[DS01] J. Dall and P. Sibani. Faster Monte Carlo simulations at low temperatures. The waiting time method. *Comp. Phys. Comm.*, 141:260–267, 2001.

[EA75] S.F. Edwards and P.W. Anderson. Theory of spin glasses. *J. Phys. F*, 5:965–974, 1975.

[EMBP$^+$09] D. El Masri, G. Brambilla, M. Pierno, G Petekidis, A. Schofield, L. Berthier, and L. Cipelletti. Dynamic light scattering measurements in the activated regime of dense colloidal hard spheres. *J. Stat. Mech.*, 100:07015, 2009.

[EMS89] M. Eigen, J. McCaskill, and P. Schuster. The molecular quasi-species. *Adv. in Chem. Phys.*, 75:149–263, 1989.

[Fel66] W. Feller. *An Introduction to Probability Theory and its Applications, vol. II*. John Wiley, New York, London, Sydney, Toronto, 1966.

[FH91] K.H. Fischer and J.A. Hertz. *Spin Glasses*. Cambridge University Press, Cambridge, 1991.

[FHS08] A. Fischer, K.H. Hoffmann, and P. Sibani. Intermittent relaxation in hierarchical energy landscapes. *Phys. Rev. E.*, 77:041120, 2008.

[Fré27] M. Fréchet. Sur la loi de probabilitè de l'écart maximum. *Ann. Soc. Polon. Math.*, 6:93, 1927.

[FT28] R.A. Fisher and L.H.C. Tippett. Limiting forms of the frequency distribution of the largest or smallest member of a sample. *Math. Proc. Camb. Phil. Soc.*, 24:180–190, 1928.

[Gal78] J. Galambos. *The Asymptotic Theory of Extreme Order Statistics*. John Wiley & Sons, New York, 1978.

[GE77] S. Gould and N. Eldredge. Punctuated equilibria: The tempo and mode of evolution reconsidered. *Paleobiology*, 3:115–151, 1977.

[Gil77] D.T. Gillespie. Exact Stochastic Simulation Of Coupled Chemical-Reactions. *J. Phys. Chem.*, 81(25):2340–2361, 1977.

[Gin86] S.L. Ginzburg. Nonergodicity and nonequilibrium character of spin glasses. *Sov. Phys. JETP*, 63:439–446, 1986.

[Gli78] N. Glick. Breaking records and breaking boards. *The American Mathematical Monthly*, 85:2–26, 1978.

[Gol89] D.E. Goldberg. *Genetic Algorithms in Search, Optimization and Machine Learning*. Addison-Wesley, Reading, MA, 1989.

[Gou02] S.J. Gould. *The Structure of Evolutionary Theory*. The Belknap Press of Harvard University Press, Cambridge, MA, London, 2002.

[Gum58] E.J. Gumbel. *Statistics of Extremes*. Columbia University Press, New York, 1958.

[HCdCJ02] M. Hall, K. Christensen, S.A. di Collobiano, and H.J. Jensen. Time-dependent extinction rate and species abundance in a tangled-nature model of biological evolution. *Phys. Rev. E*, 66:011904, 2002.

[HE10] H.J. Jensen and E. Arcaute. Complexity, collective effects and modelling of ecosystems: formation, function and stability. *Annals of the NY Acad of Sci.*, 464:2207–2217, 2010.

[HS88] K.H. Hoffmann and P. Sibani. Diffusion in hierarchies. *Phys. Rev. A*, 38:4261–4270, 1988.

[HSS97] K.H. Hoffmann, S. Schubert, and P. Sibani. Age reinitialization in spin-glass dynamics and in hierarchical relaxation models. *Europhys. Lett.*, 38:613–618, 1997.

[HW11] G.L. Hunter and E.R. Weeks. The physics of the colloidal glass transition. *Rep. Prog. Phys.*, 75:066501, 2012.

[Jen90] H.J. Jensen. Lattice gas as a model of 1/f noise. *Phys. Rev. Lett.*, 64:3103–3106, 1990.

[Jen98] H.J. Jensen. *Self-Organized Criticality. Emergent Complex Behavior in Physical and Biological Systems*. Cambridge University Press, Cambridge, 1998.

[JJS10a] D. Jones, H.J. Jensen, and P. Sibani. The tangled nature model as an evolving quasi-species model. *Ecological Modelling*, 221:400–404, 2010.

[JJS10b] D. Jones, H.J. Jensen, and P. Sibani. Tempo and mode of evolution in the tangled nature model. *Phys. Rev. E*, 82:036121, 2010.

[JKK08] T. Jörg, H.G. Katzgraber, and F. Krzakala. Behavior of Ising Spin Glasses in a Magnetic Field. *Phys. Rev. Lett.*, 100:197202, 2008.

[JN01] H.J Jensen and M. Nicodemi. Off equilibrium glassy properties of vortex creep in superconductors. *Europhys. Lett.*, 54:566–572, 2001.

[JNP+00] D.K. Jackson, M. Nicodemi, G. K. Perkins, N. A. Lindop, and H. J. Jensen. Stacking of vortices: The origin of the second peak in the magnetisation loops of high temperature superconductors. *Europhys. Lett.*, 52:210–216, 2000.

[JPS99] S.E. Jørgensen, B.C. Patten, and M. Straškraba. Ecosystems emerging: 3. Openness. *Ecological Modelling*, 117:41–64, 1999.

[JVH+98] K. Jonason, E. Vincent, J. Hammann, J-P. Bouchaud, and P. Nordblad. Memory and Chaos Effects in Spin Glasses. *Phys. Rev. Lett.*, 81:3243–3246, 1998.

[KA93] W. Kob and H.C. Andersen. Kinetic lattice-gas model of cage effects in high-density liquids and a test of mode-coupling theory of the ideal-glass transition. *Phys. Rev. E*, 48(6):4364–4377, 1993.

[Kam92] N.G. Van Kampen. *Stochastic Processes in Physics and Chemistry*. North-Holland, Amsterdam, 1992.

[KBSR10] G.G. Kenning, J. Bowen, P. Sibani, and G.F. Rodriguez. Temperature Dependence of Fluctuation Time Scales in Spin Glasses. *Phys. Rev. B*, 81:014421, 2010.

[KJ05] J. Krug and K. Jain. Breaking records in the evolutionary race. *Physica A: Statistical Mechanics and its Applications*, 358(1):1–9, 2005.

[KKM+90] C. Keller, H. Küpfer, R. Meier-Hirmer, U. Wiech, V. Selvamanickam, and K. Salama. Irreversible behaviour of oriented grained $YBa_2Cu_3O_x$. Part 2: relaxation phenomena. *Cryogenics*, 30:410–416, 1990.

[KN00] S. Kotz and S. Nadarajah. *Extreme Value Distributions: Theory and Applications*. Imperial College Press, London, 2000.

[Kra40] H.A. Kramers. Brownian motion in a field of force and the diffusion model of chemical reactions. *Physica*, 7:284–304, 1940.

[KRO06] G.G. Kenning, G.F. Rodriguez, and R. Orbach. End of Aging in a Complex System. *Phys. Rev. Lett.*, 97:057201, 2006.

[Kru07] J. Krug. Records in a changing world. *Journal of Statistical Mechanics, Theory and Experiment*, page 07001, 2007.

[KSH98] T. Klotz, S. Schubert, and K.H. Hoffmann. The state space of short-range Ising spin-glasses: the density of states. *European Physical Journal B*, 2:313–317, 1998.

[Lan74] J.-M. Lanore. Simulation de l'evolution des defauts dans un reseau par le methode de Monte-Carlo. *Radiation Effects*, 22:153–162, 1974.

[LNSB85] L. Lundgren, P. Nordblad, P. Svedlindh, and O. Beckman. Towards equilibrium in spin glasses. *J. of Appl. Phys.*, 57:3371, 1985.

[LOH+91] M. Lederman, R. Orbach, J.M. Hammann, M. Ocio, and E. Vincent. Dynamics in spin glasses. *Phys. Rev. B*, 44:7403–7412, 1991.

[LSNB83] L. Lundgren, P. Svedlindh, P. Nordblad, and O. Beckman. Dynamics of the relaxation time spectrum in a CuMn spin glass. *Phys. Rev. Lett.*, 51:911–914, 1983.

[LT94] R. Lenski and M. Travisano. Dynamics of adaptation and diversification: A 10,000-generation experiment with bacterial populations. *Proc. Natl. Acad. Sci.*, 91:6808–6814, 1994.

[MJN+95] J. Mattsson, T. Jonsson, P. Nordblad, H. Aruga Katori, and A. Ito. No Phase Transition in a Magnetic Field in the Ising Spin Glass $Fe_{0.5}Mn_{0.5}TiO_3$. *Phys. Rev. Lett.*, 74:4305–4308, 1995.

[MRR+53] N. Metropolis, A.W. Rosenbluth, M.N. Rosenbluth, A.H. Teller, and E. Teller. Equation of State Calculations by Fast Computing Machines. *J. Chem. Phys.*, 21:1087–1092, 1953.

[MSIR10] Y. Murase, T. Shimada, N. Ito, and P. A. Rikvold. Effects of demographic stochasticity on biological community assembly on evolutionary time scales. *Phys. Rev. E*, 81:041908, 2010.

[Myd93] J.A. Mydosh. *Spin glasses: an experimental introduction*. Taylor & Francis, London, Washington DC, 1993.

[NB99] M.E.J. Newman and G.T. Barkema. *Monte Carlo Methods in Statistical Physics*. Oxford University Press, Oxford, New York, 1999.

[NBW10] P. Nouvellet, J.P. Bacon, and D. Waxman. Testing the level of ant activity associated with quorum sensing: An empirical approach leading to the establishment and test of a null-model. *Journal of Theoretical Biology*, 266(4):573–583, 2010.

[NE99] M.E.J. Newman and G.J. Eble. Decline in the extinction rate and scale invariance in the fossil record. *Paleobiology*, 25:434–439, 1999.

[Nev01] V.B. Nevzorov. *Records: Mathematical Theory*. American Mathematical Society, 2001.

[NJ00] M. Nicodemi and H.J. Jensen. Second magnetisation peak relaxation in a model for vortices in superconductors. *Physica C*, 341–348:1065–66, 2000.

[NJ01a] M. Nicodemi and H.J. Jensen. Creep of superconducting vortices in the limit of vanishing temperature: A fingerprint of off equilibrium dynamics. *Phys. Rev. Lett.*, 86:4378–81, 2001.

[NJ01b] M. Nicodemi and H.J. Jensen. Off equilibrium magnetic properties in a system of repulsive particles for vortices in superconductors. *J. Phys. A*, 34:L11–L18, 2001.

[NMRP00] V. Normand, S. Muller, J.-C. Ravey, and A. Parker. Gelation kinetics of gelatin: A master curve and network modeling. *Macromolecules*, 33:1063–1071, 2000.

[NSLS86] P. Nordblad, P. Svedlindh, L. Lundgren, and L. Sandlund. Time decay of the remanent magnetization in a CuMn spin glass. *Phys. Rev. B*, 33:645–648, 1986.

[OJNS05] L.P. Oliveira, H.J. Jensen, M. Nicodemi, and P. Sibani. Record dynamics and the observed temperature plateau in the magnetic creep rate of type ii superconductors. *Phys. Rev. B*, 71:104526, 2005.

[OS85] A.T. Ogielski and D.L. Stein. Dynamics on ultrametric spaces. *Phys. Rev. Lett.*, 55:1634–1637, 1985.

[Par79] G. Parisi. Infinite number of order parameters for spin-glasses. *Phys. Rev. Lett.*, 43:1754–1756, 1979.

[Par80] G. Parisi. The order parameter for spin glasses: A function on the interval 0-1. *J. Phys. A*, 13:1101–1112, 1980.

[Par83] G. Parisi. Order parameter for spin-glasses. *Phys. Rev. Lett.*, 50: 1946–1948, 1983.

[PAS04] L.P. Oliveira P. Anderson, H.J. Jensen and P. Sibani. Evolution in complex systems. *Complexity*, 10:49–56, 2004.

[Pru11] G. Pruessner. *Self-Organised Criticality. Theory, Models and Characterisation*. Cambridge University Press, Cambridge, 2012.

[PSAA83] R.G. Palmer, D.L. Stein, E. Abrahams, and P.W. Anderson. Models of hierarchically contrained dynamics for glassy relaxation. *Phys. Rev. Lett.*, 53:1–18, 1983.

[RCF$^+$11] T.O. Richardson, K. Christensen, N.R. Franks, H.J. Jensen, and A.B. Sendova-Franks. Group dynamics and record signals in the ant *Temnothorax albipennis*. *J. R. Soc. Interface*, 8:518–528, 2011.

[Rie93] H. Rieger. Non-equilibrium dynamics and aging in the three dimensional Ising spin-glass model. *J. Phys. A*, 26:L615–L621, 1993.

[Rik07] P.A. Rikvold. Self-optimization, community stability, and fluctuations in two individual-based models of biological coevolution. *J. Math. Biol.*, 55:653–677, 2007.

[RKM$^+$86] W. Reim, R.H. Koch, A.P. Malozemoff, M.B. Ketchen, and H. Maletta. Magnetic Equilibrium Noise in Spin-Glasses: $Eu_{0.4}Sr_{0.6}S$. *Phys. Rev. Lett.*, 57:905–908, 1986.

[RKO03] G.F. Rodriguez, G.G. Kenning, and R. Orbach. Full Aging in Spin Glasses. *Phys. Rev. Lett.*, 91:037203, 2003.

[ROB87] P. Refregier, M. Ocio, and H. Bouchiat. Equilibrium Magnetic Fluctuations in Spin Glasses: Temperature Dependence and Deviation from $1/f$ Behaviour. *Europhys. Lett.*, 3:503–510, 1987.

[RRC$^+$10] T.O. Richardson, E.J.H. Robinson, K. Christensen, H.J. Jensen, N.R. Franks, and A.B. Sendova-Franks. Record Dynamics in Ants. *PLoS ONE*, 5:e9621, 03 2010.

[RS07] P.A. Rikvold and V. Sevim. Individual-based predator-prey model for biological coevolution: fluctuations, stability, and community structure. *Phys. Rev. E.*, 75:051920, 2007.

[Rud66] W. Rudin. *Real and Complex Analysis*. McGraw-Hill, New York, 1966.

[RZ03] P.A. Rikvold and R.K.P. Zia. Punctuated equilibria and 1/f noise in a biological coevolution model with individual-based dynamics. *Phys. Rev. E*, 68:031913, 2003.

[SBA98] P. Sibani, M. Brandt, and P. Alstrøm. Evolution and extinction dynamics in rugged fitness landscapes. *Int. J. Modern Phys. B*, 12:361–391, 1998.

[SC11] P. Sibani and S. Christiansen. Non-stationary aging dynamics in ant societies. *Journal of Theoretical Biology*, 282:36–40, 2011.

[Sch85] M. Schreckenberg. Long range diffusion in ultrametric spaces. *Z. Phys. B*, 60:483–488, 1985.

[Sch10] J.C. Schön. Studying the energy hypersurface of multi-minima systems --- the threshold and the lid algorithm. *Berichte der Bunsengesellschaft für physikalische Chemie*, 100:1388–1391, 2010.

[SD03] P. Sibani and J. Dall. Log-Poisson statistics and pure aging in glassy systems. *Europhys. Lett.*, 64:8–14, 2003.

[SGN+87] P. Svedlindh, P. Granberg, P. Nordblad, L. Lundgren, and H.S. Chen. Relaxation in spin glasses at weak magnetic fields. *Phys. Rev. B*, 35:268–273, 1987.

[SH89] P. Sibani and K.H. Hoffmann. Hierarchical models for aging and relaxation in spin glasses. *Phys. Rev. Lett.*, 63:2853–2856, 1989.

[SH91] P. Sibani and K.H. Hoffmann. Relaxation in complex systems: local minima and their exponents. *Europhys. Lett.*, 16:423–428, 1991.

[Sib07] P. Sibani. Linear response in aging glassy systems, intermittency and the Poisson statistics of record fluctuations. *Eur. Phys. J. B*, 58:483–491, 2007.

[Sim44] G.G. Simpson. *Tempo and Mode in Evolution.* Columbia University Press, New York, 1944.

[Sim62] H.A. Simon. The architecture of complexity. *Proc. of the American Philosophical Society*, 106:467–482, 1962.

[SJ05] P. Sibani and H.J. Jensen. Intermittency, aging and extremal fluctuations. *Europhys. Lett.*, 69:563–569, 2005.

[SK75] D. Sherrington and S. Kirkpatrick. Solvable model of a spin-glass. *Phys. Rev. Lett.*, 35:1792–1796, 1975.

[SK10] P. Sibani and G.G. Kenning. Origin of end-of-aging and subaging scaling behavior in glassy dynamics. *Phys. Rev. E*, 81:011108, 2010.

[SL93] P. Sibani and P.B. Littlewood. Slow Dynamics from Noise Adaptation. *Phys. Rev. Lett.*, 71:1482–1485, 1993.

[SM75] H. Scher and E.W. Montroll. Anomalous transit-time dispersion in amorphous solids. *Phys. Rev. B*, 12:2455–2477, 1975.

[Soo05] W.W.-H. Soon. Variable solar irradiance as a plausible agent for multidecadal variations in the Arctic-wide surface air temperature record of the past 130 years. *Geophy. Res. Lett.*, 32:L16712, 2005.

[SP99] P. Sibani and A. Pedersen. Evolution dynamics in terraced NK landscapes. *Europhys. Lett.*, 48:346–352, 1999.

[SPJ96] J.C. Schön, H. Putz, and M. Jansen. Studying the energy hypersurface of continuous systems–the threshold algorithm. *J. Phys.: Condens. Matter*, 8:143–156, 1996.

[SS94] P. Sibani and P. Schriver. Phase-structure and low-temperature dynamics of short range Ising spin glasses. *Phys. Rev. B*, 49:6667–6671, 1994.

[SSA95] P. Sibani, M. Schmidt, and P. Alstrøm. Fitness optimization and decay of the extinction rate through biological evolution. *Phys. Rev. Lett.*, 75:2055–2058, 1995.

[SSF02] P. Salamon, P. Sibani, and R. Frost. *Facts, Conjectures and Improvements for Simulated Annealing*. SIAM, Philadelphia, 2002.

[SSSA93] P. Sibani, C. Schön, P. Salamon, and J-O. Andersson. Emergent hierarchical structures in complex system dynamics. *Europhys. Lett.*, 22:479–485, 1993.

[Str78] L.C.E. Struik. *Physical Aging in Amorphous Polymers and Other Materials*. Elsevier Science, New York, 1978.

[SW83] F.H. Stillinger and T.A. Weber. Dynamics of structural transitions in liquids. *Phys. Rev. A*, 28:2408–2416, 1983.

[Tin04] M. Tinkham. *Introduction to Superconductivity*. Dover, 2004.

[Tou77] G. Toulouse. Theory of the frustration effect in spin glasses. *Communication on Physics*, 2:115–119, 1977.

[Vin91] E. Vincent. Slow dynamics in spin glasses and other complex systems. In D.H. Ryan (Editor). *Recent Progress in Random Magnets*, pp. 209–246. McGill University, Montreal, 1991.

[Wal03] D.J. Wales. *Energy Landscapes*. Cambridge Univerity Press, Cambridge, 2003.

[Wri86] S. Wright. *Sewall Wright, Evolution: Selected Papers*. University of Chicago Press, Chicago, 1986.

[You98] A.P. Young (Editor). *Spin Glasses and Random Fields*. World Scientific, Singapore, 1998.

Index